FLORA OF TROPICAL EAST AFRICA

POLYGALACEAE

JORGE PAIVA[1]

Annual or perennial herbs, shrublets or shrubs, rarely small trees or woody climbers. Leaves usually alternate, rarely opposite, fascicled or verticillate, sometimes in basal leaf-rosettes, sessile or shortly petiolate, simple, entire, usually exstipulate (stipules when present glandular or conical-truncate). Flowers hermaphrodite, zygomorphic, in 1–several-flowered, terminal, axillary or pseudolateral racemes, secondarily spiciform or corymbiform. Sepals 5, free, sometimes the 2 anterior ones joined; 3 external, 2 anterior (inferior) alike and 1 superior larger; 2 internal (lateral) ones often petaloid (wing-sepals), sometimes all nearly alike. Petals 3–5, ± joined, at least near the base, or free, subequal or unequal and the 2 lateral ones often absent or vestigial; the 2 upper ones free or joined to the lower one, usually keel-shaped (keel or carina), often fringed (crested) or appendaged at the tip. Stamens (4–)5–10, but usually 8; filaments usually united into a slit tube; anthers basifixed, introrse, monothecate. Ovary superior, (1–)2–5-locular, usually with 1 pendulous ovule per loculus [4 or (6–)8–20, very rarely ± 40 in *Xantophylleae*]; style terminal, simple; stigma with 2 spreading arms, the posterior (inferior) one papillose. Fruit a capsule, samara or drupe. Seeds often hairy, usually carunculate; endosperm mostly present, oily and proteinaceous.

A family with 4 tribes, 17–18 genera and ± 1300 species of a cosmopolitan distribution (absent in New Zealand and arctic and antarctic areas).

Of the 4 genera of the area, *Carpolobia* G.Don belongs to the tribe *Carpolobieae* B.Eriksen and the other three (*Securidaca* L., *Muraltia* DC. and *Polygala* L.) to the tribe *Polygaleae*.

1. Fruit drupaceous; petals 5; stamens 5; ovary 3-locular . . . . 1. **Carpolobia**
   Fruit capsular or samaroid; petals 3 or if 5, the 2 lateral ones
   vestigial; stamens 7–8(–10); ovary 2-locular 2
2. Fruit 1-seeded, indehiscent, samaroid; ovary with 1 abortive
   locule; upper petals free from the keel . . . . 2. **Securidaca**
   Fruit 2-seeded, dehiscent, capsular; ovary with 2 fertile
   locules; upper petals joined at the base to the keel 3
3. Fertile stamens 7; keel with a bilobed leafy crest . . . . 3. **Muraltia**
   Fertile stamens 8(–10) or 6 fertile with 2 sterile; keel with a
   fimbriate crest or plurilobed, but lobes not leafy, rarely
   absent . . . . 4. **Polygala**

1. CARPOLOBIA

G. Don, Gen. Syst. 1: 349, 370 (1831); Breteler & Smissaert-Houwing in Meded. Landb. Wagen. 77(18): 1–45 (1977)

Shrubs, small trees or lianas. Stipules usually absent, when present glandular or conical-truncate. Leaves alternate. Flowers in axillary racemes. Bracts and bracteoles small, usually long persistent. Sepals 5, unequal, the 2 internal (lateral) slightly

[1] Botanical Institute, P.O. Box 3011, 3001-455 Coimbra, Portugal

FIG. 1. *Carpolobia goetzei* – **1**, habit × 1; **2**, flower × 2; **3**, sepal × 4; **4**, upper sepal × 4; **5**, sepal × 4; **6**, petals and stamens × 2; **7**, pistil × 2; **8**, seed × 2; **9**, bud × 2. All from "J.H. & J.M.D.". Drawn by D.R. Thompson.

larger. Petals 5, joined at the base to the staminal sheath, subequal in length, slightly or shortly unequal in shape, and then the lower one keel-shaped, appendaged at the tip. Fertile stamens 5, united at the base into a sheath; staminodes 0–3. Ovary 3-locular. Fruit drupaceous, lobed, 1–3-seeded. Seeds ecarunculate, densely hairy.

A genus of four species, confined to tropical Africa.

Carpolobia goetzei *Gürke* in E.J. 28: 417 (1900); T.T.C.L.: 454 (1949); Petit in F.C.B. 7: 283 (1958); Breteler & Smissaert-Houwing in Meded. Landb. Wagen. 77(18): 30/4, map 5 (1977); K.T.S.L.: 86, fig., map (1994). Type: Tanzania, Uzaramo District, near Dar es Salaam, *Goetze* 4 (B†, holo.; K!, lecto., BM!, iso.)

Shrub or small tree up to 5 m tall; branchlets puberulous, glabrescent or glabrous. Stipules absent or visible as a dark, gland-like, glabrous spot, usually elevated or sometimes even conical-truncate, up to 1 mm high. Leaves papery to thinly coriaceous, shortly petiolate, obovate to elliptic or ovate-elliptic, (1.5–)2.5–12(–18) × (1–)1.5–6(–7.5) cm, acuminate or rounded, obtuse, base rounded or cuneate; upper surface glabrous except for the midrib, sometimes sparsely pubescent below with rather long, reddish-brown hairs on midrib often extending towards the base or leaves entirely glabrous; petiole (1–)1.5–3(–4) mm. Flowers white, cream or yellowish, with purple, red to bright pink honeyguides on tips of the petals; the corolla turns pinkish-brown after fertilization, scented, in axillary racemes (1–3 per axil) up to 3.5 cm long, 2–9-flowered; peduncle 3 mm long, pubescent; rachis pubescent; bracts and bracteoles often soon caducous, triangular, 0.5–1.5(–2) mm long, pubescent or glabrous; the basal ones often sterile and up to 3(–4) mm long. Sepals pale green, broadly ovate-elliptic, (3–)4–10 × (2–)3–6 mm, the largest up to 2 times the smaller, the upper one keel-shaped, the 2 inner ones obtuse or emarginate, glabrous or brownish pubescent outside, more densely towards the base, ciliate. Petals subequal in size, (10–)12–15(–18) × 2–4(–5) mm, the median one narrowly obovate-spathulate, gradually tapering into the claw, the inner ones oblong-elliptic, puberulous outside, ciliate. Stamens (7–)9–12 mm long; anthers 1 mm long. Fruit ovoid-subglobose, (1–)1.5(–2) cm in diameter, usually lobed, 1–3-seeded. Seeds ellipsoid, flattened, 7–9(–10) × 3–6(–7.5) mm, covered with silky rusty brownish hairs. Fig. 1, p. 2.

UGANDA. Ankole District: Ruizi River, 16 Nov. 1950 (fr.), *Jarrett* 239!; Mengo District: Entebbe, June 1905, *E. Brown* 244! & Oct. 1922 (fr.), *Maitland* 273!
KENYA. Kwale District: Shimba Hills, Mdogo forest, 5 Feb. 1953, *Drummond & Hemsley* 1135!; Kilifi District: Chasimba, 22 km SW of Kilifi, 6 Oct. 1974, *Adams* 93!
TANZANIA. Mwanza District: Rubya Forest Reserve, 3 Feb. 1960, *Carmichael* 750!; Kigoma District: Kasye Forest, 20 Mar. 1994, *Bidgood, Mbago & Vollesen* 2849!; Iringa District: Mwanihana Forest Reserve, Sange River Area, Nov. 1981, *Rodgers & Hall* 1356!; Zanzibar District: Zanzibar I., Kesmi Kazi, 30 Sept. 1960, *Faulkner* 2723!
DISTR. **U** 2, 4; **K** 5 (fide EA), 7; **T** 1, 3, 4, 6–8; **Z**; Southern Sudan, Eastern Congo-Kinshasha, Zambia, Mozambique; Madagascar
HAB. Moist and dry forest, riverine forest, wooded grassland, forest edges and thicket; 0–1400 m

SYN. *C. alba* G.Don var. *zanzibarica* Chodat in Bull. Herb. Boiss. 5: 118 (1897). Type: Tanzania, Uzaramo District: Dar es Salaam, *Kirk* 119 (K!, lecto.)
    C. conradsiana Engl., Pflanzenw. Afr. 3(1): 839 (1915); T.T.C.L.: 454 (1949); Exell in F.Z. 1, 1: 304 (1960) Type: Tanzania, Mwanza District: Ukerewe I., *Conrads* 5739 (K!, neo.; EA, iso.!)
    C. suaveolens Meikle in K.B. 1950: 337 (1951). Type: Mozambique, Lugela, Namagoa, *Faulkner* 106 (K!, holo.; BR!, FT!, P!, iso.)

USES. The fruit contains a sweet pulp that is edible, like the other three species of the genus (*C. alba* G. Don, *C. glabrescens* Hutch. & Dalz. and *C. lutea* G. Don).

2. SECURIDACA

L., [Gen. Pl. ed. 5: 316 (1754) *pro parte*] Syst. Nat. ed. 10 (2): 1155 (1759), *nom. conserv.*

Small trees or shrubs, mostly scandent or lianas. Stipules absent, when present glandular. Leaves alternate. Flowers in axillary or terminal racemes or paniculate. Bracts and bracteoles small, caducous. Sepals 5, unequal, free, the 2 internal (lateral) ones larger and petaloid (wing-sepals). Petals 3, occasionally 5, and then the 2 lateral ones vestigial, the 2 upper ones free from the lower one, which is ± keel-shaped (keel or carina), without a crest or appendages. Stamens 8, partially fused at the base into a tapering sheath. Ovary 2-locular, one abortive. Fruit a 1-locular, 1-seeded, 1-winged nut (samara), occasionally with an additional rudimentary wing. Seed without caruncle and endosperm.

A genus with ± 80 species, the majority of which occur in America, about 8 in Indo-Malaysia and 2 in Africa.

Scandent shrub or liana; leaves 5–15 × (1.5–)2.5–5 cm; inflorescence racemose or paniculate; pedicels 5–8 (–10) mm .... 1. *S. welwitschii*
Shrub or small tree; leaves (1–)2.5–5.5(–9) × 0.5–2(–2.5) cm; inflorescence racemose; pedicels 10–18(–20) mm ..... 2. *S. longepedunculata*

1. **Securidaca welwitschii** *Oliv.* in F.T.A. 1: 135 (1868); T.T.C.L.: 457 (1949); Keay, F.W.T.A., ed. 2, 1: 110 (1954); Petit, F.C.B. 7: 279 (1958); Exell in F.Z. 1(1): 305 (1960); Johnson in S. Afr. J. Bot. 53(1): 10/1A & fig. 3 (1987); K.T.S.L.: 86, map (1994). Type: Angola, Golungo Alto, Alto Queta, *Welwitsch* 994 (LISU!, lecto.; BM!, COI!, K!, isolecto.)

Scandent shrub or liana, 10–15 m tall; young branchlets minutely appressed-pubescent, soon becoming glabrous. Leaves thinly coriaceous, petiolate, broadly to oblong-elliptic, 5–15 × (1.5–)2–5.5 cm, acuminate, base cuneate to obtuse, upper surface shiny and glabrous or nearly so, glabrous below; petiole 5–8 mm, pubescent. Flowers pink or purple, sometimes variegated with white, sweet-scented, in axillary or terminal racemes or simply branched panicles 2–10(–20) cm long, often borne on short shoots which become spiny, many-flowered; peduncle and rachis pubescent; bracts and bracteoles soon caducous; pedicels 5–8(–10) mm, pubescent. Posterior sepal ovate, 4–5 × 3–4 mm, ciliate; wing-sepals suborbicular, (4–)5–9(–11) mm in diameter; anterior sepals suborbicular to broadly ovate, 4–4.5 mm in diameter. Upper petals narrowly elliptic or oblong-obovate, (4.5–)5–6(–7.5) × 2–3 mm, ciliate at the base; carina 7–10 mm × 5 mm, with a small lobed appendage 1 mm long near the apex. Staminal sheath (3–)4–8 mm long, ciliate on the upper margin. Fruit 3–5 × 0.8–2 cm, with an oblong-elliptic somewhat obliquely curved wing, with a second imperfect and rudimentary wing; nut subglobose, 8–10 mm in diameter, rugulose or smooth.

UGANDA. Bunyoro District: Masindi, Sept. 1935, *Hancock* 6A!; Busoga District: Busoga, 1904, *Dawe* 83!; Mengo District: Bukasa, Sese, 25 Feb. 1933, *A.S. Thomas* 880!
KENYA. Nandi District: S of Yala R. Bridge, 19 Apr. 1965, *Gillett* 16733!; North Kavirondo District: Kakamega Forest, near Forester's House, 15 Oct. 1953, *Drummond & Hemsley* 4766!
TANZANIA. Lushoto District: Makuyuni, 16 Dec. 1935, *Koritschoner* 1405! & 1439!
DISTR. **U** 2–4; **K** 3, 5; **T** 3, 7; West tropical Africa, Congo-Kinshasa, Angola, Zambia
HAB. Upland evergreen forest; 1100–1650 m

NOTE. In Tanzania the root is reputed to be toxic.

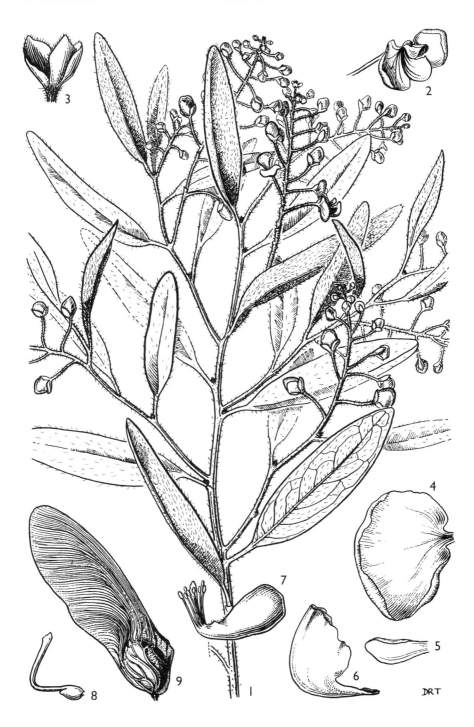

Fig. 2. *Securidaca longepedunculata* – **1**, habit × 1; **2**, flower × 2; **3**, calyx × 4; **4**, wing sepal × 4; **5–6**, petals × 4; **7**, pistil & stamens × 4; **8**, pistil × 3; **9** fruit × 1.5. All from Liebenberg 1577. Drawn by D.R. Thompson.

2. **Securidaca longepedunculata** *Fresen.* in Mus. Senckenb. 2: 275 (1837); Oliv., F.T.A. 1: 134 (1868); T.T.C.L.: 457 (1949); Milne-Redh. in Mem. N.Y. Bot. Gard. 8(3): 219 (1953); Keay, F.W.T.A., ed. 2, 1: 110 (1954); Petit, F.C.B. 7: 280 (1958); Exell in F.Z. 1(1): 305, t. 55/A (1960); Johnson in S. Afr. J. Bot. 53(1): 5/1B & fig. 2 (1987); K.T.S.L.: 86, fig., map (1994); Gilbert in Fl. Eth. 2, 1: 177, t. 24, 1 (2000). Type: Ethiopia, without locality, *Rüppel* 40 (FR!, lecto. & isolecto.)

Small tree or shrub up to 6 m tall; young branchlets softly pubescent, becoming glabrous. Leaves petiolate, oblong-ovate, oblong-elliptic to linear-lanceolate, (1–)2.5–5.5(–9) × 0.5–2(–2.5) cm, coriaceous, obtuse to rounded, base cuneate; upper surface glabrous or slightly pubescent below; petiole 2–10 mm, pubescent. Flowers pink or purple, sometimes variegated with white, sweet-scented, in axillary or terminal racemes 3–15 cm long, often borne on short shoots which become spiny, many-flowered; peduncle and rachis pubescent; bracts and bracteoles soon caducous; pedicels 10–18(–20) mm, pubescent. Posterior sepal ovate, 4–6 × 3–5 mm, ciliate; wing-sepals suborbicular, (3–)5–9.5 mm in diameter; anterior sepals suborbicular to broadly ovate, 2.5–4.5 mm in diameter. Upper petals narrowly elliptic or oblong-obovate, 5–7.5 × 2.5–4 mm, ciliate towards the base; carina 8–10 × 5 mm, with a small lobed appendage, 1 mm long, near the apex. Staminal sheath 6–7 mm long, ciliate on the upper margin. Fruit 5–7 × 1–2 cm, with an oblong-elliptic somewhat obliquely curved wing, with a second imperfect and rudimentary wing; nut subglobose, 8–10 mm in diameter, rugulose or smooth. Fig. 2, p. 5.

UGANDA. West Nile District: Koboko, Mar. 1939, *Hazel* 721!; Ankole District: Ruizi R., 15 Nov. 1950, *Jarrett* 281!; Mengo District: Wabusana, Bulemezi, 2 June 1935, *A.S. Thomas* 1270!
KENYA. Embu District: Itabua, 29 Nov. 1932, *Graham* 2250!; North Kavirondo District: Webuye Falls, Mar. 1967, *Tweedie* 3435!; Kwale District: N of Mrima Hill, 29 Jan. 1961, *Greenway* 9006!
TANZANIA. Mbulu District: Babati, Sigino Valley, 19 May 1988, *Ruffo & Sigara* 3039!; Lushoto District: between Soni and Mombo, 1 June 1971, *Shabani* 708!; Morogoro District: Ruvu Forest Reserve, 4 Jan. 1977, *Magogo* 754!
DISTR. **U** 1–4; **K** 4, 5, 7; **T** 1–8; widespread in tropical Africa, from Senegal to Ethiopia, to Namibia and South Africa
HAB. Wooded and bushed grassland, scattered tree grassland, open forest or woodland, secondary bushland; 0–1700 m

SYN. *Lophostylis angustifolia* Hochst. in Flora 25: 231 (1842). Type: Ethiopia, Tigre, near Ferrefaram, *Schimper* s.n. (TUB!, holo.; GH!, iso.)
 L. oblongifolia Hochst. in Flora 25: 231 (1842). Type: Ethiopia, Tigre, near Tacaze, *Schimper* 771 (TUB!, holo.; K!, MO!, NY!, iso.)
 L. pallida Klotzsch in Peters, Nat. Reise Mossamb. Bot. 1: 115, t. 22 (1861). Type: Mozambique, Sena, *Peters* s.n. (B†, holo.)
 S. longepedunculata Fresen. var. *angustifolia* Robyns in Bull. Soc. Roy. Bot. Belg. 60: 90 (1927). Type: Congo-Kinshasha, Mpala, *Delevoy* 412 (BR!, lecto.)

USES. In Tanzania the root is used against roundworm, to expel intestinal parasites in children and as cathartic against constipation. Sometimes grown as an ornamental.

3. **MURALTIA**

DC., Prodr. 1: 335 (1824); Levyns in S. Afr. J. Sci., suppl. 2: 1–247 (1954)

Ericoid shrubs or shrublets. Stipules absent. Leaves alternate or fascicled, usually apiculate or pungent. Flowers axillary, sessile or pedicellate. Bracts and bracteoles small, usually persistent. Sepals 5, usually subequal, sometimes the 2 internal (lateral) ones slightly or sometimes distictly larger (wings). Petals usually 3, the 2 upper (lateral) ones joined at the base to the lower one (keel or carina), keel-shaped, dinstinctly differentiated into claw and limb, with a 2-lobed, leafy

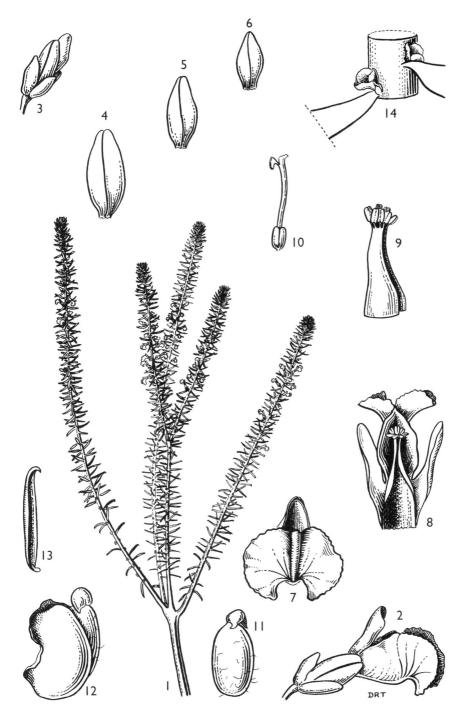

FIG. 3. *Muraltia flanaganii* – **1**, habit ×; **2**, flower × 6; **3**, bud × 8; **4–6**, sepal × 8; **7**, carina and crest, front view × 8; **8**, petals and stamens × 8; **9**, stamens × 8; **10**, pistil × 8; **11**, seed × 6; **12**, capsule and seed × 6; **13**, leaf × 4; **14**, bracteoles × 8. All from *Johnston* 20. Drawn by D.R. Thompson

appendage (crest) attached at or close the tip. Fertile stamens 7, united at the base into a split tube, adhering below to the petals. Ovary 2-locular. Fruit a flattened capsule, usually 4-horned at the apex, 2-seeded. Seeds with a caruncle, pubescent or glabrous.

A genus with about 115 species confined to South Africa, except *M. flanaganii.*

Muraltia flanaganii *Bolus* in J.B. 34: 17 (1896); Milne-Redh. in Mem. N.Y. Bot. Gard. 8 (3): 220 (1935); T.T.C.L.: 455 (1949); Levyns in S. Afr. J. Sci., suppl. 2: 34, t. 7/A-J (1954); Exell in F.Z. 1(1): 307, t. 55/B (1960). Type: South Africa, Natal, Mont aux Sources, *Flanagan* 2020 (K!, holo.)

Shrub up to 1 m tall; branches angled, glabrous or minutely pubescent. Leaves subsessile, linear or acicular, 3–8 × 0.5–1 mm, obtuse or apiculate with a needle-like point, usually revolute, glabrous or scaberulous on the margins. Flowers white, pink or purple, usually solitary; bracts and bracteoles ovate, 0.5 × 0.8 mm, obtuse-rounded at the apex; pedicels 1 mm. Sepals subequal, the 3 outer ones slightly smaller than the 2 inner ones, ovate, 2–3 × 1.5–2 mm, glabrous or ciliate. Upper petals oblong-elliptic, 4 mm long, obtuse, shortly clawed; carina 3.5 mm long, usually with the claw slightly longer than the limb; crest 1.5 mm long. Stamen-tube 3–3.5 mm long; anthers subglobose. Capsule flattened broadly elliptic in outline, 3 × 2.5 mm, with the apical projections very short, not horned, glabrous. Seeds ellipsoid, 1.5 × 0.8 mm, sparsely pubescent; caruncle 0.5 mm long, with very short appendages. Fig. 3, p. 7.

TANZANIA. Moshi District: Kilimanjaro, above rain gauge, 16 July 1993, *Grimshaw* 93411!; Njombe
   District: Kipengere Mts, 13 Jan. 1957, *Richards* 7735!; Rungwe District: Poroto Mts, Livingstone
   Forest Reserve, 28 Sept. 1970, *Thulin & Mhoro* 1242!
DISTR. **T** 2, 7; Malawi, Mozambique, Zimbabwe and South Africa
HAB. Upland grassland and moorland; 1800–3000 m

SYN. *M. fernandi* Chodat in Mitt. Bot. Mus. Univ. Zür. 76: 612 (1916); T.T.C.L.: 455 (1949). Type:
   Malawi, Mlanje Plateau, *McClounie* s.n. (K!, syn.)
   M. nyassana Mildbr. *in sched.* Based on: Tanzania, Rungwe District: Kyimbila, *Stolz* 1036 (K!)

4. **POLYGALA**

L., Sp. Pl. 2: 701 (1753) & Gen. Pl. ed. 5: 315 (1754); Paiva in Fontqueria 50: I–VI,
1–346, tab. 1–52 (1998)

Annual or perennial herbs, shrublets, shrubs, rarely small trees (or woody climbers in Madagascar). Leaves usually alternate, rarely opposite or verticillate, sometimes in basal leaf-rosettes, sessile or shortly petiolate, simple, entire, usually linear to lanceolate, exstipulate. Flowers in several-flowered, terminal, axillary or pseudolateral racemes, secondarily spiciform or corymbiform, rarely solitary. Sepals 5, unequal, 3 external, persistent (caducous in some American subgenera), 2 anterior (inferior) alike, free, sometimes united and 1 superior larger; 2 internal (lateral) ones larger and petaloid (wing-sepals). Petals 3(5), the 2 lateral ones usually absent or vestigial, the 2 upper ones joined at the base to the lower one (keel or carina), keel-shaped and often fringed (crested); crest fimbriate or plurilobed, rarely without the crest. Stamens usually 8 (exceptionally 9–10), united into a slit tube, sometimes 6 fertile and 2 staminodes. Ovary 2-locular, sessile, sometimes shortly stalked; stigma 2-branched or 2-lobed. Fruit a capsule, usually flattened, 2-locular, 2-seeded, usually winged. Seeds often hairy, with a silky indumentum, rarely glabrous or with glochidiate hairs, usually with a 3-lobed or 3-branched caruncle, often with 3 membranous appendages.

A cosmopolititan genus with about 750 species, widespread mainly in tropical and subtropical regions. The genus is divided into several subgenera, but the African species belong to the subgenus *Polygala*, with the exception of three species from NW Africa [(*P. munbyana* Boiss. (Algeria and Morocco), *P. webbiana* Coss. (Morocco) and *P. balansae* Coss. (Morocco)], which belong to subgenus *Chamaebuxus* (DC.) Rchb. Species 1–6 belong to the section *Timutua* DC., 7 to *Leptaleae* (Chodat) Paiva, 8–32 to *Blepharidium* DC., 33–42 to *Tetrasepalae* (Chodat) Paiva, 43 to *Conospermae* (Chodat) Paiva, 44–47 to *Chloropterae* (Chodat) Paiva and 48–50 to *Megatropis* Paiva.

1. Anterior sepals united for at least half their length, usually
    almost to the apex 2
   Anterior sepals free or only slightly joined at the base 11
2. Fertile stamens 6 3
   Fertile stamens 8 8
3. Wing sepals caducous in fruit; flowers in terminal racemes 4
   Wing sepals persistent in fruit; flowers in lateral and
    terminal racemes or only in lateral or only in
    terminal racemes 5
4. Shrublet, rarely annual herb, 50–100 cm tall; wing sepals
    6–9 × 2.5–4 mm; sterile stamens hairy; seeds ellipsoid,
    4–6 × 2–2.5 mm ....   39. *P. acicularis*
   Annual or perennial herb 30–70 cm tall; wing sepals
    3–3.5 × 1–1.3 mm; stamens glabrous; seeds conical,
    2.5–3 × 1–1.2 mm ....   40. *P. amboniensis*
5. Seeds conical, with the convex base completely covered
    by glands and a pyramidal caruncle completely
    covered by hairs; wing sepals 3-veined from the base
    and only anastomosing venation near the base .....   42. *P. muratii*
   Seeds ellipsoid or obovoid-ellipsoid, with a glandless
    base and a subglobose or subovoid caruncle not
    covered by hairs; wing sepals 3–5-veined from the
    base and with anastomosing venation   6
6. Simple or few-branched annual herb, 30–100 cm tall;
    leaves 25–60 mm long; flowers in terminal racemes,
    10–30 cm long ....   41. *P. conosperma*
   Shrublet or perennial herb with woody base, sending
    up annual shoots 5–25 cm long; leaves 5–25 mm long,
    mucronate; flowers solitary or in terminal or lateral
    racemes, up to 6 cm long   7
7. Leaves linear, with a needle-like point, up to 1 mm
    wide; flowers solitary or in few-flowered, lateral
    racemes, up to 1.5 cm long ....   37. *P. gossweileri*
   Leaves oblanceolate to obovate-linear, exceptionally
    linear, (2–)3–6 mm wide; flowers in terminal or many-
    flowered lateral racemes, up to 6 cm long ....   38. *P. luteo-viridis*
8. Carina with a crest   9
   Carina without a crest, exceptionally with a vestigial one   10
9. Flowers in terminal racemes, up to 20 cm long; leaves
    lanceolate to narrowly lanceolate, 5–30 × 2–7 mm,
    pointed; wing sepals 5–8 × 4–5.5 mm, blue or
    greenish with purple-brown veins ....   33. *P. stenopetala*
   Flowers in lateral racemes up to 10–15 cm long; leaves
    oblong-elliptic to obovate, 10–20 × 5–10 mm, rounded
    to obtuse; wing sepals 5 × 3.5 mm, greenish with green
    veins ....   34. *P. multifurcata*

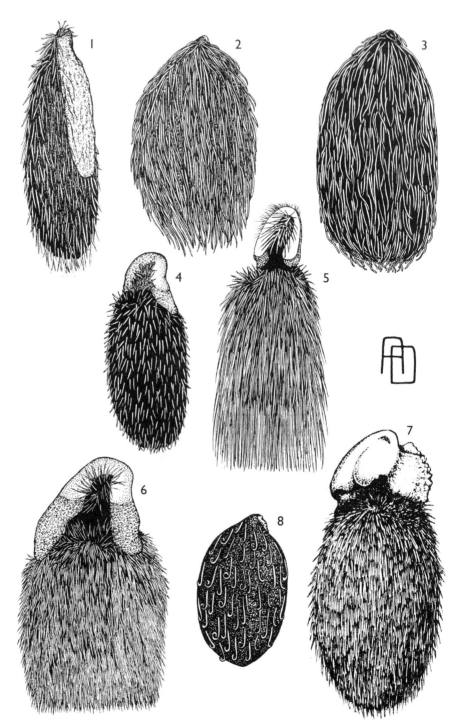

FIG. 4. *Polygala* seeds – **1**, *P. paniculata* × 40; **2**, *P. spicata* × 60; **3**, *P. capillaris* × 60; **4**, *P. myriantha* × 40; **5**, *P. erioptera* × 20; **6**, *P. sadebeckiana* × 20; **7**, *P. humifusa* × 50; **8**, *P. africana* × 60. 1 from *Drummond & Hemsley* 2115; 2 from *Baum* 315; 3 from *Drege* s.n.; 4 from *Rand* 139; 5 from *Vogel* 4; 6 from *Torre* 5011; 7 from *Richards* 7697; 8 from *Richards* 5632. Drawn by Ann Davies.

FIG. 5. *Polygala* seeds – **1**, *P. gillettii* × 20; **2**, *P. senensis* × 15; **3**, *P. sphenoptera* × 20; **4**, *P. vittata* × 25; **5**, *P. arenaria* × 20; **6**, *P. welwitschii* subsp. *pygmaea* × 30; **7**, *P. melitioides* × 30. 1 from *Gillett* 13084; 2 from *Phipps* 171; 3 from *Torre* 1463; 4 from *Maas Geesteranus* 6162; 5 from *Espirito Santo* 1349; 6 from *Drummond & Hemsley* 4708; 7 from *Milne-Redhead & Taylor* 105/1. Drawn by Ann Davies.

10. Annual herb up to 50(–70) cm tall, with ridged arcuate-
　　ascending stems; wing sepals (3.5–)4.5–6 × 2.5–3 mm,
　　with 3–5 greenish veins . . . .　36. *P. xanthina*
　　Annual herb up to 100(–150) cm tall, with slender erect
　　stems; wing sepals (2–)3–5 × (0.8–)1.8–3 mm, with
　　(3–)4–5 blue veins . . . .　35. *P. petitiana*
11. Seeds conical, with the convex base completely covered
　　by glands . . . .　43. *P. irregularis*
　　Seeds ellipsoid, ovoid, obovoid, cylindric or globose 12
12. Seeds without caruncle or with a very reduced caruncle
　　(usually reduced to the membranaceous appendages);
　　capsules ellipsoid, obovoid-ellipsoid or subglobose,
　　unwinged 13
　　Seeds with a conspicuous caruncle; capsules flattened-
　　ovoid or flattened-ellipsoid, ± winged 20
13. Seeds with appressed hairs; caruncle absent or reduced
　　to the membranaceous appendages 14
　　Seeds with glochidiate hairs or glabrous; caruncle absent 17
14. Seeds with the caruncle reduced to the membranous
　　appendages; leaves alternate, but the lower ones in
　　1–2 verticels of 4–5 leaves . . . .　1. *P. paniculata*
　　Seeds without caruncle; all leaves alternate 15
15. Flowers in lax or dense racemes; wing sepals (1.5–)1.7–2
　　× 0.7–1 mm; pedicels usually up to 0.5 mm long; seeds
　　(0.7–)1 mm long . . . .　3. *P. capillaris*
　　Flowers always in dense racemes; wing sepals 2–3 ×
　　1–1.7 mm; pedicels 0.5–1 mm long; seeds 0.5–0.8 mm
　　long 16
16. Flowers pale greenish-yellow or whitish; capsule stipitate;
　　seeds with brownish, rarely whitish hairs . . . .　2. *P. spicata*
　　Flowers pinkish; capsule sessile; seeds with white hairs . .　4. *P. sansibarensis*
17. Herbs (15–)20–60 cm tall; wing sepals (1.5–)1.7–3 ×
　　0.7–1.5 mm 18
　　Herbs 8–18(–25) cm tall; wing sepals 1–1.5 × 0.8–1.2 mm 19
18. Wing sepals (2–)2.5–3 × 1–1.5 mm . . . .　4. *P. sansibarensis*
　　Wing sepals (1.5–)1.7–2 × 0.7–1 mm . . . .　3. *P. capillaris*
19. Seeds with glochidiate hairs . . . .　5. *P. africana*
　　Seeds glabrous or with a few and rare glochidiate hairs　6. *P. afra*
20. Bracts and bracteoles caducous 21
　　Bracts and bracteoles persistent 25
21. Posterior sepals almost twice as long as the anterior
　　ones; wing sepals 2–3 times as long as wide, narrowly
　　elliptic to elliptic . . . .　7. *P. myriantha*
　　Posterior sepals almost as long as the anterior ones;
　　wing sepals 1.5 times as long as wide, broadly elliptic
　　to suborbicular or obovate 22
22. Racemes terminal or exceptionally with a few additional
　　lateral ones, many-flowered; shrubs or shrublets,
　　rarely perennial herbs 23
　　Racemes lateral, few-flowered (1–15 flowers), annual or
　　perennial herbs, rarely shrublets 50
23. Shrubs or shrublets; wing sepals suborbicular or ±
　　oblong-elliptic, (6–)9–15 mm wide, deep purple to
　　pale lilac . . . .　48. *P. virgata*
　　Perennial herbs; wing sepals oblong-elliptic, 3–5 mm
　　wide, pink, white or greenish 24

FIG. 6. *Polygala* seeds – **1**, *P. albida* subsp. *albida* × 20; **2**, *P. albida* subsp. *stanleyana* × 20; **3**, *P. schweinfurthii* × 20; **4**, *P. persicariifolia* × 20; **5**, *P. kasikensis* × 20; **6**, *P. ruwenzoriensis* × 20. 1 from *Pawek* 4693; 2 from *Richards* 16444; 3 from *Evrard* 1801; 4 from *Rogers* 10928; 5 from *De Giorgi* s.n.; 6 from *de Wilde* 290. Drawn by Ann Davies.

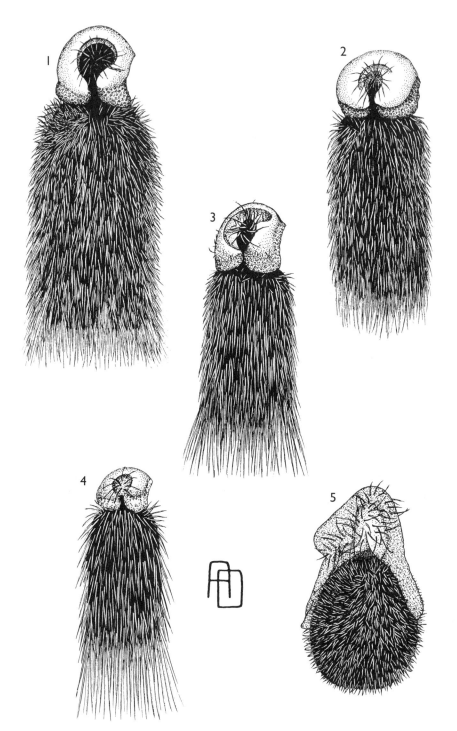

FIG. 7. *Polygala* seeds – **1**, *P. exelliana* × 20; **2**, *P. bakeriana* × 20; **3**, *P. usafuensis* × 20; **4**, *P. sparsiflora* × 20; **5**, *P. ohlendorfiana* × 20. 1 from *Robyns* 3946; 2 from *Elskens* 31; 3 from *Goetze* 1032; 4 from *Morton & Gledhill* 2977; 5 from *Torre & Correia* 15644. Drawn by Ann Davies.

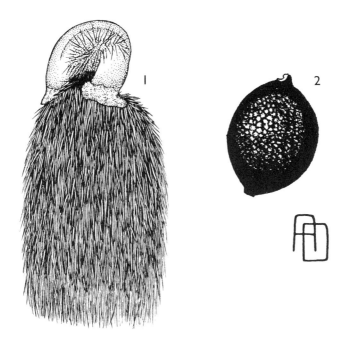

FIG. 8. *Polygala* seeds – **1**, *P. stenopetala* × 20, from *Torre* 5025; **2**, *P. afra* × 50, from *Greenway & Polhill* 11666. Drawn by Ann Davies.

24. Alpine herb up to 18 cm tall; stems crisped-pubescent; wing sepals 5–5.5 × 3–3.5 mm; caruncular appendages very long, up to half the length of the seed . . . .   50. *P. steudneri*
     Upland herb, 30–50 cm tall; stems sparsely pubescent to glabrous; wing sepals 7–8.5 × 3.5–5 mm; caruncular appendages almost inconspicuous, 0.1–0.2 mm long .   49. *P. abyssinica*
25. Racemes terminal and lateral or only terminal 26
     Racemes only lateral 42
26. Racemes terminal and lateral shortly exceeding the upper leaves 27
     Racemes terminal, sometimes with some lateral, much exceeding the upper leaves 32
27. Capsules not ciliate; wing sepals 2–4 × 2–3 mm 28
     Capsules ciliate; wing sepals (3.5–)4–9 × (2.5–)3–6.5 mm 29
28. Leaves elliptic to broadly elliptic, 10–30(–40) × (3–)5–12 (–20) mm; seeds ovoid to subglobose, 1–1.8(–2) × 0.6–0.8(–1) mm, glabrous, rarely puberulous; caruncular appendages hiding the caruncle . . . .   19. *P. melilotoides*
     Leaves lanceolate to oblong-elliptic, 15–30(–60) × 1–10 mm; seeds ellipsoid to subglobose, 1.5–2.3 × 0.7–1 mm, silky pubescent (rarely puberulous); caruncular appendages shorter than the caruncle . . .   18. *P. welwitschii*
29. Herb of sympodial growth, secondary flowering branchlets much exceeding the primary terminal racemes . . . .   17. *P. arenaria*
     Herb unbranched or with the secondary flowering branchlets usually shorter than the terminal primary racemes 30

30. Terminal and lateral racemes similar in length and
    number of flowers; wing sepals suborbicular-ovate to
    broadly elliptic, glabrous or ciliate . . . .   22. *P. persicariifolia*
    Terminal racemes longer and more flowered than the
    lateral ones; wing sepals obliquely ovate-elliptic or
    ovate, ± pubescent, at least at the base 31
31. Bracts glabrous or laxly ciliate; venation of the wing
    sepals prominent; style 3–4 mm long . . . .   20. *P. albida*
    Bracts ciliate; venation of the wing sepals ± inconspicuous;
    style 6–7 mm long . . . .   21. *P. schweinfurthii*
32. Leaves up to 20 cm long, not darkish when dried 33
    Leaves up to 10 cm long, darkish when dried 35
33. Bracts and bracteoles subequal, 2.5–3.5 mm wide;
    racemes congested; wing sepals 14–14.5 mm long . . .   23. *P. kasikensis*
    Bracts and bracteoles unequal, 0.5–1.5 mm wide;
    racemes lax or ± congested; wing sepals shorter than
    14 mm 34
34. Bracts caudate (base ovate with thread-like apex); seed
    hairs long (± 2 mm long), exceeding the base of the
    seed . . . .   25. *P. macrostigma*
    Bracts acuminate or caudate-acuminate (without a
    thread-like apex), ovate-lanceolate to lanceolate;
    seed hairs shorter (± 0.5 mm long), never exceeding
    the base of the seed . . . .   24. *P. ruwenzoriensis*
35. Bracts 7–12 mm long; wing sepals 11–13 × 10–12.5 mm . .   24. *P. ruwenzoriensis*
    Bracts 1–4 mm long; wing sepals up to 11 × 10 mm 36
36. Perennial herb with prostrate stems, 5–20 cm long,
    spreading from a woody rootstock 37
    Shrublets or annual or perennial erect herbs, 20–400 cm
    tall 38
37. Leaves glabrous (except when young, ± pubescent),
    narrowly oblong-elliptic, 2–6 mm wide, cuneate at
    the base; seed ellipsoid; caruncle symmetric, kidney-
    shaped . . . .   28. *P. nyikensis*
    Leaves pilose or crisped-pubescent, ovate, ovate-
    lanceolate to suborbicular, 3–14 mm wide, rounded
    or obtuse to very slightly cordate at the base; caruncle
    strongly asymmetric, comma-shaped . . . .   32. *ohlendorfiana*
38. Shrublet or perennial herb up to 4 m tall; pedicels
    9–13 mm long; capsule 6–9 × 4–5 mm, shiny . . . .   26. *P. exelliana*
    Annual or perennial herbs or small shrublets up to 2.5 m
    tall; pedicels 1.5–7 mm long; capsule up to 6 × 4 mm,
    dull 39
39. Caruncle conical, much longer than wide, 1–1.5 mm
    long, without membranaceous appendages . . . .   31. *P. nambalensis*
    Caruncle kidney-shaped, as long as wide, up to 0.8 mm
    long, membranous appendages scarcely developed 40
40. Inflorescences branched at the base; flowers brownish
    when dried; pedicels 5–7 mm long; capsule brownish
    when dried . . . .   27. *P. bakeriana*
    Inflorescences not branched, or with only a few
    branches; flowers pinkish or yellowish when dried;
    pedicels 2–3.5 mm long; capsule yellowish when
    dried 41

FIG. 9. *Polygala* seeds – **1**, *P. petitiana* subsp. *petitiana* var. *petitiana* × 20; **2**, *P. petitiana* subsp. *petitiana* var. *abercornensis* × 20; **3**, *P. petitiana* subsp. *parviflora* × 20; **4**, *P. xanthina* × 20; **5**, *P. luteo-viridis* × 20. 1 from *Gilbert* 2; 2 from *Richards* 4679; 3 from *Buchman* 366; 4 from *Lewalle* 932; 5 from *Michel* 5312. Drawn by Ann Davies.

41. Robust herb, 1.5–2 m tall, densely patent-pubescent;
    leaves up to 70 × 10 mm, the ones at the base of the
    stems shorter and wider; racemes straight; wing
    sepals 6–8 × 4–6 mm; carina pubescent on the back;
    seeds (3–)3.5–5 × 1–1.5 mm . . . .   29. *P. usafuensis*
    Slender herb, 0.3–1 m tall, scarcely crisped-pubescent,
    rarely patent-pubescent; leaves up to 30 × 1.5 mm, ±
    similar; racemes usually flexuous; wing sepals 4.5–5.5
    × 3–3.5 mm (var. *ukirensis*); carina glabrous; seeds
    3–3.5 × 1–1.3 mm . . . .   30. *P. sparsiflora*
42. Racemes shorter or a few longer than the proximal leaf,
    few-flowered; wing sepals elliptic to oblong obovate 43
    Racemes much longer than the proximal leaf, usually
    many-flowered; wing sepals suborbicular or obovate-
    orbicular 46
43. Wing sepals narrower than the capsules, with a
    prominent green stripe down the midrib, densely
    pubescent . . . .   8. *P. erioptera*
    Wing sepals as wide as the capsules, without a
    prominent green stripe down the centre, glabrous or
    sparsely pubescent 44
44. Membranous caruncular appendages (at least one)
    almost as long as the length of the seed . . . .   11. *P. gillettii*
    Membranous caruncular appendages absent or shorter
    than the length of the seed 45
45. Annual creeping herb; leaves suborbicular to broadly
    elliptic, 3.5–6 × 3–4.5 mm, undulate on the margin;
    wing sepals 3.5–4 × 1.7–2 mm . . . .   10. *P. humifusa*
    Shrublets or ± erect herbs; leaves elliptic to obovate,
    10–50(–90) × (5–)10–25(–40) mm, not undulate;
    wing sepals (4.5–)5–7 × 2.5–3.5(–4) mm . . . .   9. *P. sadebeckiana*
46. Wing sepals conspicuously and densely reticulate; two
    membranous caruncular appendages large, almost as
    long as the seed 47
    Wing sepals inconspicuously and laxly reticulate;
    membranous caruncular appendages much shorter
    than the length of the seed 48
47. Perennial or occasionally annual herb, up to 60 cm tall,
    densely crisped-pubescent, sometimes patent-
    pubescent; two membranous caruncular appendages
    usually humped at the apex of the seed . . . .   13. *P. senensis*
    Shrublet or perennial herb, up to 150 cm tall,
    densely patent-pubescent; membranous caruncular
    appendages never humped at the apex of the seed   12. *P. kilimandjarica*
48. Wing sepals obliquely obovate-orbicular, with two red
    or purple-brownish stripes down the centre, which
    delimit a middle fringe . . . .   16. *P. vittata*
    Wing sepals suborbicular, concolorous, sometimes with
    purple margin when dried 49

FIG. 10. *Polygala* seeds – **1**, *P. acicularis* × 12; **2**, *P. amboniensis* × 24; **3**, *P. conosperma* × 24; **4**, *P. irregularis* × 24; **5**, *P. transvaalensis* × 20; **6**, *P. goetzei* × 20. 1 from *Le Testu* 2929; 2 from *Hildebrandt* 1204; 3 from *Hildebrandt* 1925; 4 from *Bartha* 72; 5 from *Lebrun* 9796; 6 from *Torre & Correia* 10129. Drawn by Ann Davies.

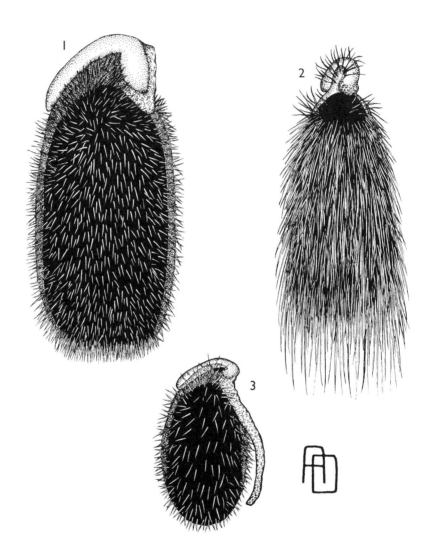

FIG . 11. *Polygala* seeds – **1**, *P. virgata* var. *decora* × 20; **2**, *P. abyssinica* × 20; **3**, *P. steudneri* × 20. 1 from *Leach & Cannell* 14282; 2 from *Lynes* 69; 3 from *Mooney* 6123. Drawn by Ann Davies.

51. Perennial erect herb, 30–40 cm tall; secondary venation
 and reticulation of the leaves conspicuous . . . .   47. *P. goetzei*
 Annual or perennial sprawling herbs or subshrubs
 5–20 cm tall; secondary venation and reticulation of
 the leaves inconspicuous 52
52. Wing sepals 4.5–6 × 2.5–3 mm; capsule 4.5–6 mm in
 diameter . . . .   46. *P. transvaalensis*
 Wing sepals 3–4 × 1.2–1.6 mm; capsule 2.5–3.5 mm in
 diameter . . . .   45. *P. meonantha*

1. **Polygala paniculata** *L.*, Syst. Nat., ed. 10, 2: 1154 (1759); Paiva in Fontqueria 50: 154, t. 25/a (1998); Bernardi in Cavanillesia Altera 1: 13, tabs. 1–3 (2000). Type: Jamaica, *Browne* s.n. (LINN!, holo.)

Annual herb, 15–50 cm tall, often branched from the base. Stems slender, glandular-pubescent. Leaves alternate, but the lower ones in 1–2 verticels of 4–5 leaves, petiolate, linear to linear-lanceolate, 10–25 × 1–2 mm, acute, base cuneate, glabrous; petiole 0.5 mm, glabrous or slightly glandular-pubescent. Flowers pink, in terminal racemes (occasionally some lateral) 2–5(–9.5) cm long; rachis glabrous; bracts linear, 0.6–0.8 mm long, glabrous and caducous; bracteoles linear, 0.2 mm long, glabrous and caducous; pedicels 0.7–1 mm, glabrous. Posterior sepal linear, 1.2–1.3 mm long, glabrous; wing sepals ovate-elliptic, 2–2.5 × 0.7–1 mm, glabrous; anterior sepals linear, 1.2 mm long, glabrous, free. Upper petals obovate, 2.5 mm long; carina 1–1.5 mm long; crest 1 mm long, 3–4-fimbriate. Stamens 8. Capsule obliquely obovate-elliptic in outline, 1.75 × 1–1.3 mm, unwinged, glabrous. Seeds ellipsoid, 1.2–1.5 × 0.5 mm, sparsely pubescent; caruncle reduced to the caruncular appendages up to as long as half of the length of the seed. Fig. 4: 1, p. 10.

KENYA. Kwale District: Shimba Hills, 3.5 km from Kwale to Tanga, 15 Nov. 1968, *Magogo & Estes* 1231!
TANZANIA. Lushoto District: Soni-Bumbuli road, 0.8 km N of Kwehangala Pt., 11 May 1953, *Drummond & Hemsley* 2515!; Tanga District: Magila, near Muheza, 5 Feb. 1970, *Archbold* 1260!
DISTR. **K** 7; **T** 3; widespread in the tropical regions of the World; introduced from South America
HAB. Cultivated land, grassland, path- and road-sides; 350–1700 m

SYN. *Polygala amaniensis* Chodat in E.J. 48(1–2): 311 (1912), *nom. nud.* Based on: Tanzania, Lushoto District: Amani, *col.*? (B†)

USES. This species is used as a medicinal plant against snake bites and blenorrhagias (which is the reason why this is the species of *Polygala* with the largest distribution).
NOTE. C. van der Meijden [Polygalaceae in Flora Malesiana, ser. 1, 10(3): 455–539 (1988)] and L. Bernardi [Cavanillesia altera 1: I-VIII; 1–456 (2000)] considered *P. fernandesiana* Paiva, from Equatorial Africa (Chad, Nigeria, Cameroon and Central-African Republic) as a synonyms of *P. paniculata* L. Indeed they look like the same species but *P. fernandesiana* has smaller capsules (1.2–1.5 × 0.75 mm) and seeds (0.7–0.8 × 0.4 mm) than *P. paniculata* (1.7–2 × 1.1–1.2 mm and 1.3–1.5 × 0.5 mm) and the seeds of the former do not have caruncular appendages while the latter has two caruncular appendages up to as long as half of the length of the seed. *P. paniculata* is usually a ruderal plant in Africa, while *P. fernandesiana* grows in grassland.

2. **Polygala spicata** *Chodat* in Mém. Soc. Phys. Hist. Nat. Genève 31 (2): 221, tab. 23/36–37 (1893); Petit in F.C.B. 7: 270 (1958); Exell in F.Z. 1, 1: 327, t. 57/14 (1960); Paiva in Fontqueria 50: 157, t. 25/c (1998). Type: Angola, Lopolo, *Welwitsch* 1027 (B†, holo.; LISU!, lecto.; BM!, COI!, K!, isolecto.)

Annual herb, (15–)20–40 cm tall. Stems simple or a little branched, slender, wiry, glabrous. Leaves alternate, scarcely developed and sometimes a little broader near the base of the stem, subsessile, linear to lorate, 5–8 × 0.2–1 mm, acute, base cuneate, glabrous. Flowers pale greenish-yellow or whitish, in dense terminal racemes

2.5–15(–20) cm long; rachis glabrous; bracts linear, 1–1.5 mm long, glabrous, caducous; bracteoles linear, 0.7–0.8 mm long, glabrous, caducous; pedicels 0.5–1 mm, glabrous. Posterior sepal linear, 1.5–1.8 × 1 mm, glabrous; wing sepals obovate-elliptic to broadly elliptic, 2–2.5 × 1.5–1.8 mm, glabrous; anterior sepals linear, 1–1.3 × 0.3–0.5 mm, glabrous, free. Upper petals broadly elliptic, 1.5–1.8 mm long; carina 1.5 mm long, 0.7–0.8 mm high; crest 0.5–0.7 mm long, 3–4-fimbriate. Stamens 8. Capsule stipitate, suborbicular in outline, 1 mm diameter, unwinged, glabrous. Seeds ellipsoid, 0.5–0.8 × 0.4–0.5 mm, with brownish, rarely whitish hairs; caruncle absent. Fig. 4: 2, p. 10.

UGANDA. Masaka District: NW side of Lake Nabugabo, Aug. 1935, *Chandler* 1356! & idem, 6 Oct. 1953, *Drummond & Hemsley* 4624! & Lake Kayonje, Nov. 1961, *Rose* 10250!
TANZANIA. Iringa District: Mufindi, Ngowasi Lake, 22 Mar. 1962, *Polhill & Paulo* 1830a! & 30 km on Sao Hill–Mbeya road, 30 Mar. 1988, *Bidgood et al.* 808!; Njombe District: Njombe, 2 Jan. 1932, *Lynes* 75b!
DISTR. U 4; T 7; SE and Central tropical and subtropical Africa
HAB. Wet grassland and marshy or boggy ground; 1000–2000 m

3. **Polygala capillaris** *Harv.* in Harv. & Sond., Fl. Cap. 1: 93 (1860); Petit in F.C.B. 7: 269 (1958); Exell in F.Z. 1, 1: 329, t. 57/15 (1960); Paiva in Fontqueria 50: 158, t. 7/c; t. 25/f (1998); Gilbert in Fl. Eth. 2, 1: 188, t. 24.4/15–18 (2000). Type: South Africa, between the Omsamculo and Omcomas Rs., *Drège* s.n. (K!, holo.; BM!, CGE!, E!, LD!, P!, W!, iso.)

Annual herb, (15–)20–50 cm tall. Stems simple, little-branched or much-branched, filiform and glabrous, rarely with a few glands. Leaves alternate, subsessile, linear to lorate, 3–15 × 0.2–1 mm, acute, base cuneate, glabrous. Flowers pink, purplish or greenish-white, in lax or dense terminal racemes (occasionally some lateral) 6–10(–20) cm long; rachis glabrous; bracts linear, 0.6–0.8 mm long, glabrous, soon caducous; bracteoles linear, 0.3–0.5 mm long, glabrous, soon caducous; pedicels 0.2–0.5 mm, glabrous. Posterior sepal linear, 1–1.5 mm long, glabrous; wing sepals elliptic or oblanceolate, (1.5–)1.7–2 × 0.7–1 mm, glabrous; anterior sepals linear, 0.7–1 mm long, glabrous, free. Upper petals lanceolate or elliptic, 1.5–2 mm long; carina 1.2–1.5 mm long; crest 0.5 mm long, ± fimbriate. Stamens 8. Capsule elliptic to suborbicular in outline, 1–1.5 × 1–1.3 mm, unwinged, glabrous. Seeds ellipsoid, (0.7–)1 × 0.5–0.6 mm, with minute, white appressed hairs or rarely with some glochidiate hairs; caruncle absent.

subsp. **capillaris**

Stems simple or little-branched, without or with very few glands. Fig. 4: 3, p. 10.

UGANDA. West Nile District: Koboko, Sept. 1938, *Hazel* 680!; Teso District: Soroti, Sept. 1954, *Lund* 471!; Masaka District: Bukakata, June 1925, *Maitland* 806!
KENYA. North Kavirondo District: near Bungoma, Sept. 1967, *Tweedie* 3490!
TANZANIA. Bukoba District: Bukoba, July 1931, *Haarer* 2048!; Lushoto District: Korogwe, 16 Aug. 1967, *Archbold* 930!; Iringa District: 12 km W of Sao Hill on road to Mbeya, 9 Jan. 1975, *Brummitt & Polhill* 13640!
DISTR. U 1–4; K 5; T 1, 3, 7, 8; widespread in tropical and subtropical Africa and Madagascar
HAB. Seasonally wet grassland, edges of swamps and lakes; 300–2100 m

SYN. *Polygala bukobensis* Gürke in P.O.A. C: 233 (1895). Type: Tanzania, Bukoba District: Bukoba, *Gürke* 3821 (B†, holo.; K!, iso.)
    P. volkensii sensu De Wildem. in B.J.B.B. 5: 291 (1919), *non* Gürke (1895)
    P. chevalieri sensu A.Chev. & Jacq.-Fél. in A. Chev., Fl. Afr. Oc. Franç.: 262 (1938), *pro parte quoad specim. Lecomte, non* Chodat (1907)
    P. capillaris Harv. var. *tukpwoensis* E.M.A.Petit in B.J.B.B. 25: 333 (1955) & in F.C.B. 7: 269, fig 7J (1958). Type: Congo-Kinshasha, Ubangi-Uele, Tukpwo, *Gérard* 1730 (BR!, holo.; B!, BM!, K!, P!, iso.)

P. capillaris Harv. var. *bosoboloensis* E.M.A.Petit in B.J.B.B. 26: 261 (1956). Type: Congo-Kinshasha, Ubangi-Uele, Bosobolo, *Evrard* 1320 (BR!, holo.)
    Polygala filicaulis sensu F.P.U.: 52 (1962), *non* Baill.

NOTE. This species is morphologically very variable, but the plants from West tropical Africa can be distinguished [subsp. *perrottetiana* (Paiva) Paiva in Fontqueria 50: 161, tab. 25/g (1998)] in having much-branched stems and sessile glands.

4. **Polygala sansibarensis** *Gürke* in P.O.A. C: 233 (1895); U.O.P.Z.: 419 (1949); Paiva in Fontqueria 50: 161, t. 25/i (1998). Type: Tanzania, Zanzibar Island, Kokotoni, *Hildebrandt* 1125 (B†, holo.; BM!, K!, lecto.; BM!, W!, isolecto)

Annual herb, (20–)40–60 cm tall. Stems branched in upper half, cylindric and glabrous. Leaves alternate, subsessile, linear to linear-lanceolate, 7–10 × 0.5–0.6 mm, acute, base cuneate, glabrous. Flowers pinkish, in dense terminal racemes 4–7 cm long; rachis glabrous; bracts linear-lanceolate, 0.7–0.9 mm long, glabrous, caducous; bracteoles linear, 0.4–0.5 mm long, glabrous, caducous; pedicels 0.2–1 mm, glabrous. Posterior sepal lanceolate, 1.5 mm long, glabrous; wing sepals obovate, (2–)2.5–3 × 1–1.5 mm, glabrous; anterior sepals linear, 0.7–1 mm long, glabrous, free. Upper petals obovate-elliptic, 1.8–2 mm long; carina 1–1.5 mm long; crest 0.4–0.5 mm long, 7–8-fimbriate. Stamens 8. Capsule sessile, suborbicular in outline, 1–1.3 mm diameter, unwinged, glabrous. Seeds ellipsoid-ovoid, 0.6–0.7 × 0.4–0.5 mm, with white, wavy or straight (rarely glochidiate) hairs; caruncle absent.

KENYA. Kilifi District: Mida, Oct. 1965, *Tweedie* 3181! Kwale District: near Kaya Lunguma, 20 Aug. 1994, *Luke & Gray* 4054; and Nr Ramisi, 18 Oct. 2000, *2nd Herbarium Techniques Course* 075
TANZANIA. Pangani District: Ushongo Mabaoni, Mwera, 26 June 1956, *Tanner* 3029!; Rufiji District: Mafia I., Marimbani-Adani, 9 Aug. 1937, *Greenway* 5023!; Zanzibar District: Zanzibar I., 1908, *Last* s.n.!
DISTR. **K** 7; **T** 3, 6; **Z**; NE Mozambique
HAB. Wet places and seasonally wet grassland not far from the coast; 0–10 m (–140 m near Lunguma)

NOTE. The plants from Mafia Island have smaller flowers than the typical form, and glochidiate hairs.

5. **Polygala africana** *Chodat* in Mém. Soc. Phys. Hist. Nat. Genève 31 (2): 168, tab. 21/20–21 (1893); Petit in F.C.B. 7: 268, fig. 7, I (1958); Exell in F.Z. 1, 1: 328, t. 57/13, t. 58B (1960); Paiva in Fontqueria 50: 161, t. 7/a-b; t. 9; t. 17/c; t. 25/j (1998). Type: Angola, Cuanza Norte, Pungo Andongo, *Welwitsch* 1009 [1109] (B†, holo.; LISU!, lecto.; BM!, isolecto.)

Annual herb, 8–15(–25) cm tall. Stems often much branched, filiform and glabrous. Leaves alternate, subsessile, linear-filiform, 2–7 × 0.2–0.5 mm, acute and often uncinate, base cuneate, glabrous. Flowers pink, rarely whitish, in dense, terminal racemes 1–2 cm long; rachis glabrous; bracts linear, 0.5–0.8 mm long, glabrous, soon caducous; bracteoles linear, 0.2–0.5 mm long, glabrous, soon caducous; pedicels 0.2–0.3 mm, glabrous. Posterior sepal linear-elliptic, 0.7–0.8 mm long, glabrous; wing sepals suborbicular, 1–1.5 × 1–1.2 mm, glabrous; anterior sepals linear, 0.5–0.6 mm long, glabrous, free. Upper petals oblong-lanceolate, 0.8–1 mm long; carina 1 mm long; crest 0.5 mm long, ± fimbriate. Stamens 8. Capsule suborbicular in outline, 1–1.2 mm diameter, unwinged, glabrous. Seeds subglobose to ellipsoid, 0.6 × 0.5–0.6 mm, with minute, glochidiate hairs; caruncle absent. Fig. 4: 8, p. 10.

TANZANIA. Ufipa District: 8 km on on Mkasama road from Namanyere–Chala road, 6 May 1997, *Bidgood et al.* 3777! & 20 km on Mkasama road from Namanyere–Chala road, 6 May 1997, *Bidgood et al.* 3784! & S of Sikonge, July 1979, *Acres* 99!

DISTR. **T** 4; tropical Africa S of Equator
HAB. Seasonally inundated grassland; ± 1525 m

SYN. *Polygala capillaris sensu* Harv. in Harv. & Sond., Fl. Cap. 1: 93 (1860), *pro parte quoad specim.*
 Burke & Zeyher, non E.Mey. ex Harv. (1860)
 P. micrantha sensu Oliv., F.T.A.1: 131 (1868), *pro parte quoad specim. Welwitsch, non* Perr. &
 Guill. (1831)

6. **Polygala afra** *Paiva* in Fontqueria 50: 162, t. 6/c; t. 27; t. 25/k (1998). Type: Tanzania, Dodoma District, Itigi Station, on the road to Chunya, *Greenway & Polhill* 11666 (K!, holo.)

Annual herb, 10–18 cm tall. Stems ± branched, filiform and glabrous. Leaves alternate, subsessile, linear-filiform, 4–10 × 0.5–0.8 mm, acute, glabrous. Flowers pink, in dense, terminal racemes 1–2 cm long; rachis glabrous; bracts linear-lanceolate, ± 0.7 mm long, glabrous, caducous; bracteoles linear, 0.3–0.4 mm long, glabrous, caducous; pedicels 0.2–0.3 mm, glabrous. Posterior sepal linear-elliptic, ± 0.7 mm long, glabrous; wing sepals elliptic-suborbicular, 1.2–1.5 × 0.9–1.1 mm, glabrous; anterior sepals linear, 0.4–0.5 mm long, glabrous, free. Upper petals obovate-lanceolate, 1–1.2 mm long; carina 1 mm long; crest 0.4–0.5 mm long, 2–3-fimbriate. Stamens 8. Capsule suborbicular in outline, 0.7–0.8 mm diameter, unwinged, glabrous. Seeds ellipsoid-subglobose, 0.4–0.5 × 0.3–0.4 mm, glabrous or with a few and rare glochidiate hairs; caruncle absent. Fig. 8: 2, p. 15.

TANZANIA. Dodoma District, Itigi Station, on the road to Chunya, 20 Apr. 1964, *Greenway & Polhill* 11666!
DISTR. **T** 5; only known from the type locality
HAB. Wet grassland; ± 1380 m

7. **Polygala myriantha** *Chodat* in E.J. 48: 321 (1912); Petit in F.C.B. 7: 267, fig. 7, H (1958); Exell in F.Z. 1, 1: 327, t. 57/12 (1960); U.K.W.F. ed. 2: 55 (1994); Paiva in Fontqueria 50: 165, t. 28/b (1998); Gilbert in Fl. Eth. 2, 1: 188 (2000). Type: Cameroon, *Ledermann* 1797 (B†, syn.); *ibidem, Ledermann* 15732 (B†, syn.); *ibidem, Ledermann* 16011 (B†, syn.)

Annual herb, 15–50 cm tall. Stems slender, often branched from near the base, crisped-pubescent. Leaves alternate, shortly petiolate, linear, linear-lanceolate to linear-elliptic, 7–40 × 1–7 mm, mucronate, base cuneate, sparsely pubescent to glabrous; petiole 0.5–1 mm, glabrous or slightly pubescent. Flowers mauve or lilac (drying yellowish-white), in elongated terminal or lateral racemes 3–14 cm long; rachis crisped-pubescent to almost glabrous; bracts linear-lanceolate, 1–1.8 mm long, ciliolate, caducous; bracteoles linear, 0.7–1 mm long, ciliolate, caducous; pedicels 0.5–1 mm, crisped-pubescent. Posterior sepal linear, 1.7–2 mm long, with an acute and recurved apex, sparsely crisped-pubescent, ciliolate; wing sepals narrowly elliptic to elliptic, 2.5–3 × 1–1.2 mm, ciliolate; anterior sepals linear, 1–1.3 mm long, ciliolate, free. Upper petals obliquely oblong-obovate, 1–1.5 × 0.5–0.8 mm; carina (1–)1.5–1.8 mm long; crest 0.8–1 mm long, fimbriate. Stamens 8. Capsule oblong-obovate in outline, 1.5–2 × 1–1.2 mm, margin very narrowly winged, glabrous. Seeds cylindric, (1.2–)1.5–2 × 0.5–0.6 mm, with very short ± appressed hairs; caruncle ± conical, 0.2–0.3 mm long, with very short appendages. Fig. 4: 4, p. 10.

UGANDA. Masaka District: 1.5 km from Katera on Katera–Kiebbe road, 1 Oct. 1953, *Drummond & Hemsley* 4528!; Mengo District: Jumba, May 1917, *Dümmer* 3200!
KENYA. Uasin Gishu District: Kipkarren, Dec. 1931, *Brodhurst Hill* 641! & Chemase, 16 Sep. 1960, *Tallantire* 621!

Tanzania. Ufipa District: Sumbawanga, 12 km on Tatanda–Sumbawanga road, 26 Apr. 1997, *Bidgood, Sitoni, Vollesen & Whitehouse* 3372!; Njombe District: Mukumburu, 56 July 1956, *Milne-Redhead & Taylor* 10745!; Songea District: 12 km W of Songea by Kimarampaka stream, 19 Mar. 1956, *Milne-Redhead & Taylor* 9252!

Distr. **U** 4; **K** 3; **T** 4, 7, 8; tropical Africa from Nigeria to Ethiopia and S to Angola and Zimbabwe

Hab. Grassland, *Brachystegia* and secondary woodland; 850–1650 m

Syn. *Polygala kisantuensis* Chodat in Bull. Soc. Bot. Genève, Sér. 2, 5: 189 (1913). Type: Congo-Kinshasha, Bas-Congo, Kisantu, *Gillett* 1014 (BR!, holo.; BM!, iso.)
   P. kisantuensis Chodat var. *tenuifolia* Chodat in Bull. Soc. Bot. Genève, Sér. 2, 5: 190 (1913). Type: Congo-Kinshasha, Bas-Congo,Thysville, *Bequaert* 7742 (BR!, holo.)

8. **Polygala erioptera** *DC.*, Prodr. 1: 326 (1824); Petit in F.C.B. 7: 255 (1958); Exell in F.Z. 1, 1: 316 (1960); Thulin, Fl. Somalia 1: 84 (1993); U.K.W.F. ed. 2: 56 (1994); Paiva in Fontqueria 50: 169, t. 28/c (1998); Gilbert in Fl. Eth. 2, 1: 182, fig. 24.2/16, 17 (2000). Type: Senegal, *Bacle* s.n. (G-DC!, lecto.)

Annual, rarely perennial herb or shrublet, 5–50 cm tall. Stems usually much branched from near the base, with arcuate-ascending branches, densely crisped-pubescent or, occasionally, patent-pubescent, rarely glabrescent. Leaves alternate, petiolate, linear oblong-elliptic to narrowly oblong-oblanceolate or oblong-linear, 7–30(–40) × 1.5–6 mm, rounded, sometimes emarginate, base cuneate, crisped-pubescent, rarely almost glabrous; petiole 1–1.5 mm, crisped-pubescent. Flowers greenish-white, usually pink-tipped, in short few-flowered lateral racemes up to ± 1 cm long or solitary; rachis crisped-pubescent; bracts linear, 0.7–1 mm long, pubescent, persistent; bracteoles linear, 0.5–0.7 mm long, pubescent, persistent; pedicels 1–2 mm, crisped-pubescent. Posterior sepal keel-shaped, 2–3 mm long, crisped-pubescent; wing sepals obliquely elliptic, (3.5–)4.5–5 × (1.5–)2–2.5 mm, usually with a prominent green stripe down the midrib, densely pubescent; anterior sepals linear, 1.5–2.5 × 0.5–0.8 mm, pubescent, free. Upper petals obliquely and narrowly obovate-elliptic, 2.5–3 × 0.7–1 mm; carina 3–3.5 mm long; crest ± 1 mm long, fimbriate. Stamens 8. Capsule oblong-obovate to oblong-elliptic in outline, 3–4 × 2.5–3 mm, ciliate and very narrowly winged or almost unwinged, pubescent. Seeds flattened-cylindric, 2.5–3.5 × 1.2–1.3 mm, brown-sericeous; caruncle sharply conical, 0.5–0.8 mm long, without appendages. Fig. 4: 5, p. 10.

Uganda. West Nile District: Jonam country, 19 Feb. 1969, *Lye* 2199!; Toro District: Buranga–Bwenamule road, Dec. 1925, *Maitland* 1082!; 0.8 km W of Kibimba swamp on Jinja–Busia road, 25 May 1953, *Wood* 742!
Kenya. Northern Frontier District: 1 km S of Archer's Post, 27 Mar. 1978, *Gilbert* 5042!; Fort Hall District: Thika N, Nairobi–Fort Hall road, 11 July 1971, *Kabuye* 365!; Teita District: 17 km on Sala Gate–Sobo Rocks road, 29 Dec. 1966, *Greenway & Kanuri* 12909!
Tanzania. Musoma District: Seronera, Serengeti, 2 Apr. 1961, *Greenway* 9953!; Ufipa District: Ruwa Valley, 16 Mar. 1952, *Siame* 170!; Iringa District: Kidatu, 20 Mar. 1971, *Thulin* 764!; Zanzibar District: Zanzibar I., Fumba, 19 Mar. 1964, *Faulkner* 3362!
Distr.**U** 1–3; **K** 1–7; **T** 1–7; **Z**; widespread throughout Africa (except Cape and NW); Arabian Peninsula, extending eastward to India
Hab. Various types of open bushland, woodland and grassland, often on rocky ground or lava scree, also ruderal and as a weed; 0–2050 m

Syn. *Polygala paniculata* Forssk., Fl. Aegypt.-Arab.: 117 (1775), *nom. nud., non* L. (1759)
   P. linearis R.Br. in Salt, Voy. Abyss., App. 4: 65 (1814), *nom. nud., non* Lagasca (1816)
   P. oligantha A.Rich., Tent. Fl. Abyss. 1: 38 (1847). Type: Ethiopia, Choho, *Petit* s.n. (P!, holo.)
   P. retusa Hochst. in F.T.A.1: 129 (1868), *nom. illeg.*
   P. nubica Hochst. in F.T.A. 1: 129 (1868), *nom. illeg.*
   P. triflora sensu Oliv., F.T.A. 1: 129 (1868), *non* L. (1753)
   P. erioptera DC. var. *abyssinica* Chodat in Mém. Soc. Phys. Hist. Nat. Genève 31 (2): 344 (1893). Type: Ethiopia, Mt Ghedem, *Schweinfurth & Riva* 125 (FR!, holo.; K!, iso.)

P. petraea Chodat in Mém. Soc. Phys. Hist. Nat. Genève 31 (2): 346 (1893). Type: Kenya, Kitui District: Ukamba, *Hildebrandt* 2784 (B†, holo.; M! lecto., K!, W!, isolecto.)
P. erioptera DC. subsp. *petraea* (Chodat) Paiva in Fontqueria 50: 172 (1998)

NOTE. *Polygala erioptera* DC. is widespread in several types of habitats from Africa and Arabia to tropical Asia. So it is one of the most variable and polymorphic species of the genus, varying from small ephemeral annuals to woody shrubs in East Africa (around Lake Turkana); from sparsely pubescent or glabrescent to densely grey-hairy; and from wings acuminate with prominent central stripe to obtuse without stripe. These variations are so continuous that is impossible to justify keeping the infra-specific taxa recognised by some authors, including myself!

9. **Polygala sadebeckiana** *Gürke* in P.O.A. C: 233 (1895); U.O.P.Z.: 419 (1949); Exell in F.Z. 1, 1: 316, t. 56/9 (1960); Thulin, Fl. Somalia 1: 85 (1993); U.K.W.F. ed. 2: 56 (1994); Paiva in Fontqueria 50: 173, t. 28/e (1998); Gilbert in Fl. Eth. 2, 1: 183, fig. 24.2/18–19 (2000). Type: Tanzania, Uzaramo, *Stuhlmann* 6711 (B†, syn.; BM!, lecto.; W!, isolecto.)

Perennial (rarely annual) herb or shrublet, 10–30 cm tall, sometimes sending up annual shoots after burning. Stems erect, simple or ± branched, crisped- to patent-pubescent. Leaves alternate, shortly petiolate, elliptic to obovate, 10–50(–90) × (5–)10–25(–40) mm, rounded to subacute, base cuneate to rounded, sparsely to densely pubescent, mainly on the veins and margin; petiole 0.5 mm, pubescent. Flowers greenish-white, in ± congested lateral racemes 1–2.5 cm long; rachis pubescent; bracts and brateoles similar, linear-lanceolate, 1–1.3 mm long, pubescent, persistent; pedicels 1.5–2.5 mm, pubescent. Posterior sepal keel-shaped, 2.5–3.5 mm long, pubescent; wing sepals obliquely elliptic, (4.5–)5–7 × 2.5–3.5(–4) mm, pubescent to glabrous; anterior sepals keel-shaped, 2–2.5 mm long, pubescent, free. Upper petals obliquely obovate-elliptic, 2–2.25 mm long; carina 3.5–4 mm long; crest ± 1.5 mm long, fimbriate. Stamens 8. Capsule obovate to suborbicular in outline, (3–)4–6.5 × 3.5–5 mm, ciliolate and broadly winged (wing ± 0.5 mm wide). Seeds ovate or elliptic in outline, 2.5–4 × 1.5–2 mm, sericeous; caruncle 0.5–0.7 mm long, with three appendages 0.7–2(–2.5) mm long. Fig. 4: 6, p. 10.

UGANDA. West Nile District: Terego, Apr. 1938, *Hazel* 514!; Kigezi District: Ruhinda, Jan. 1951, *Purseglove* 3532!; Mengo District: Lwala, Bulemezi, June 1925, *Maitland* 906!
KENYA. Northern Frontier District: 50 km NE of Habaswein, 27 Apr. 1978, *Gilbert & Thulin* 1124!; South Nyeri District: Mathera, Cratiki, near Sagana R., 16, May 1957, *Kibui* 17!; Kwale District: Shimba Hills, Kwale–Mombasa road, 5 May 1968, *Magogo & Glover* 996!
TANZANIA. Pangani District: Utupa, Mseko, Mwera, 21 Aug. 1956, *Tanner* 3092!; Uzaramo District: Ruvu, 28 July 1969, *Bally* 479!; Kilwa District: Kingupira Forest, 12 Jan. 1977, *Vollesen* 4315!; Pemba: Kiwani, 19 Dec. 1930, *Greenway* 2762!
DISTR. **U** 1–4; **K** 1, 3,4, 6,7; **T** 1–3, 6–8; **P**; Sudan, Ethiopia, Somalia, Malawi and Mozambique
HAB. Riverine forest and bushland, forest margin, grassland, often on 'black cotton' soils; 10–2500 m

SYN. *Polygala schimperi* Chodat in Mém. Soc. Phys. Hist. Nat. Genève 31 (2): 349, t. 28/11 (1893), non C.Presl (1845), nec Hassk. (1864). Type: Ethiopia, Sanka Berr, *Schimper* 1224 (P!, lecto.; E!, K!, isolecto.)
   P. maxima Gürke in P.O.A. C: 233 (1895); T.T.C.L.: 456 (1949). Type: Tanzania, Moshi District: Marangu, *Volkens* 2323 (B†, holo.; E!, lecto.)
   P. polygoniflora Chodat in J.B. 34: 200 (1896). Type: Malawi, Mlanje, *Scott Elliot* 8670 (BM!, holo.; K!, iso.)
   P. sadebeckiana Gürke var. *minor* Chodat in Bull. Herb. Boiss. 4(12): 906 (1896). Type: Uganda, Ruwenzori Mts, *Scott Elliot* s.n. (B†, holo.)
   P. vatkeana Exell in J.B. 70: 184 (1932); T.T.C.L.: 456 (1949); Petit in F.C.B. 7: 254, fig. 7/C (1958); Paiva in Fontqueria 50: 173, t. 28/f (1998). Type: Ethiopia, Sanka Berr, *Schimper* 1224 (P!, lecto.; E!, K!, isolecto.)
   P. kassasii Chrtek in Preslia 48(1): 83 (1976). Type: Sudan, Kassala Province, near Gedarif, *Kassas et al.* 319 (PRC!, holo.; K!, iso.)

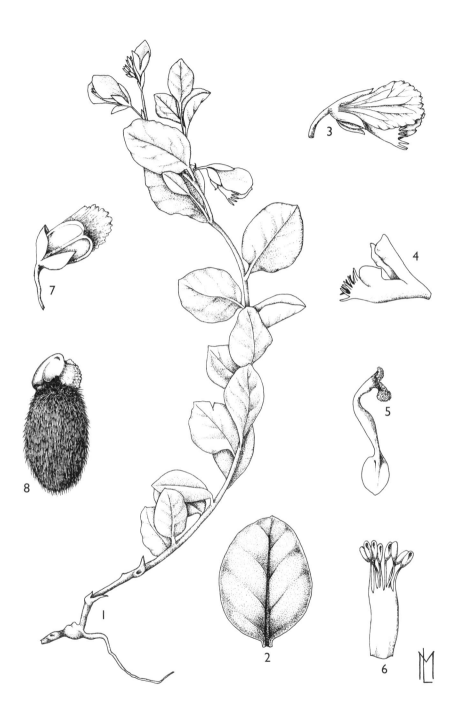

FIG. 12. *Polygala humifusa* – **1**, habit × 6 ; **2**, leaf × 12; **3**, flower × 12; **4**, carina and upper petals × 12; **5**, pistil × 25; **6**, stamens × 25; **7**, capsule and sepals × 12; **8**, seed × 50. All from *Richards* 7697. Drawn by M. Lameiras.

NOTE. The type [Tanzania, Usagara, Madessa, *Stuhmann* 8112 (B†, holo.)] of *P. cryptantha* Chodat [in Engl. E.J. 48(1–2): 311 (1912); T.T.C.L.: 456 (1949); Paiva in Fontqueria 50: 307 (1998)] has disappeared and there are no duplicates. It is almost certainly *P. sadebeckiana* Gürke.

The type of *P. erlangeri* Gürke (*Ellenbeck* 1984, from Ethiopia) has been destroyed at B. So I do not know what *P. erlangeri* is; but I do know that it cannot be in subsection *Eriopterae* (= *Blepharidium*), where Chodat puts it, because of the original description of the wings; it is in subsection *Tinctorieae* or *Asiaticae*. The material from Somalia and Kenya which Thulin in Fl. Somalia 1: 84 (1993) calls *P. erlangeri* is a different taxon (*P. kajii* Paiva), occurring in Somalia, Kenya and Tanzania, and not in Ethiopia. The type of *P. erlangeri* mentioned by Thulin, loc. cit., and by Gilbert in Fl. Eth. 2,1: 180 (2000) is probably either *P. obtusissima* or *P. sadebeckiana*.

10. **Polygala humifusa** *Paiva* in Fontqueria 50: 174, t. 6/a; t. 28/h; t. 31 (1998). Type: Tanzania, Njombe District: Elton (Kitulo) Plateau, Ndumbi R., *Richards* 7697 (K!, holo.; COI!, iso.)

Annual creeping herb up to 7 cm tall. Stems slender, branched towards the base, crisped-pubescent. Leaves alternate, petiolate, suborbicular to broadly elliptic, 3.5–6 × 3–4.5 mm, undulate on the margin, rounded at the apex and base, glabrous; petiole 0.5–0.8 mm. Flowers lilac, in 2–5-flowered lateral racemes up to 1.5 cm long, rachis crisped-pubescent, rarely solitary; bracts ovate-lanceolate, 0.7–1 mm long, persistent; bracteoles linear, 0.5 mm long, persistent; pedicels 1.5–2 mm, glabrous. Posterior sepal ovate-lanceolate, 1.7–2 × 0.8–1 mm, glabrous; wing sepals obovate-elliptic, 3.5–4 × 1.7–2 mm, margin undulate, glabrous; anterior sepals elliptic-lanceolate, 1.5 × 0.7 mm, glabrous, free. Upper petals obovate, 3–3.5 × 1 mm; carina 3 × 1.5 mm; crest fimbriate, 0.7–1 mm long. Stamens 8. Capsule elliptic-obovate in outline, 2.5 × 1.7–2 mm, narrowly winged (wing ± 0.2 mm wide). Seeds ellipsoid, 1.5 × 0.7 mm, sericeous, caruncle 0.2 mm long, without appendages. Fig. 4: 7, p. 10 & Fig. 12, p. 27.

TANZANIA. Njombe District: Elton Plateau, Ndumbi R., 11 Jan. 1957, *Richards* 7697!
DISTR. **T** 7; only known from the type
HAB. Swampy wet ground, peaty soil; 2250 m

11. **Polygala gillettii** *Paiva* in Fontqueria 50: 174, t. 30/b; t. 33 (1998). Type: Kenya, Northern Frontier District: Dandu, *Gillett* 13084 (K!, holo.; EA!, iso.)

Annual or perennial herb up to 20 cm tall. Stems slender, branched, crisped-pubescent. Leaves alternate, petiolate, elliptic to obovate, 25–40 × 10–18 mm, slightly apiculate and mucronate, base cuneate, sparsely crisped-pubescent above, mainly towards the margin, crisped-pubescent beneath, sometimes later glabrous; petiole 3–3.5 mm, crisped-pubescent. Flowers greenish, in lateral and axillary dense racemes up to 1.5 cm long, rachis densely crisped-pubescent; bracts lanceolate, 1–1.3 mm long, persistent; bracteoles linear, 0.7 mm long, persistent; pedicels 1.5–2 mm, sparsely crisped-pubescent. Posterior sepal lanceolate, 2–2.5 mm long, sparsely minutely crisped-pubescent; wing sepals elliptic, 4–4.5 × 2.5 mm, sparsely crisped-pubescent or glabrous; anterior sepals elliptic-lanceolate, 1.5 mm long, glabrous, free. Upper petals obovate, 1.7–2 mm; carina 3–3.5 × 1.5 mm; crest minute, ± 0.2 mm long, 2–branched. Stamens 8. Capsule orbicular in outline, 4.5–5 mm diameter, symmetric and deeply bilobate, clearly broader than the wing sepals, cross-veined-winged (wing 1–1.5 mm wide), crisped-pubescent, mainly towards the margin of the wing. Seeds ellipsoid, 3–3.5 × 1.2–1.5 mm, white-sericeous; caruncle asymmetric, with three appendages, one almost as long as the length of the seed (2.5–3 mm long) and the other two shorter, 0.7–1 mm long. Fig. 5: 1, p. 11 & Fig. 13, p. 29.

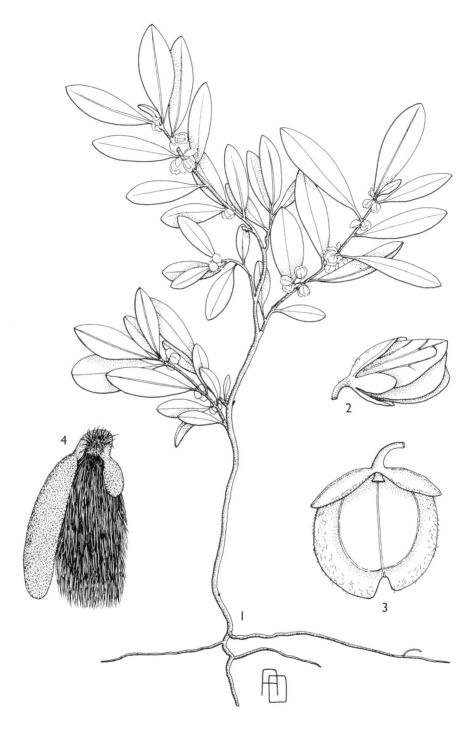

FIG. 13. *Polygala gillettii* – **1**, habit × 1 ; **2**, flower × 12; **3**, capsule and anterior sepals × 9; **4**, seed × 20. All from *Gillett et al.* 13084. Drawn by Ann Davies.

KENYA. Northern Frontier District: Dandu, 6 May 1952, *Gillett* 13084!
DISTR. **K** 1; only known from the type
HAB. Open *Acacia-Commiphora* bushland on 'black cotton' soil; ± 750 m

12. **Polygala kilimandjarica** *Chodat* in Mém. Soc. Phys. Hist. Nat. Genève 31 (2): 320, t. 27/10 (1893); T.T.C.L.: 456 (1949); Paiva in Fontqueria 50: 177 (1998). Type: Kenya, Teita District, Maungu, *Johnston* 1884 (B†, holo.; K!, lecto, BM!, isolecto.)

Perennial herb or shrublet up to 1.5 m tall. Stems spreading from a woody rootstock, branched, densely patent-pubescent. Leaves alternate, subsessile (petiole up to 0.1 mm, patent-pubescent), elliptic to obovate-elliptic, 15–40 × 5–15 mm, rounded to obtuse or sometimes emarginate, base cuneate, pubescent on both sides. Flowers pinkish, in lateral racemes 2–8(–11) cm long, rachis patent-pubescent; bracts linear, 1.2–1.5 mm long, pubescent, persistent; bracteoles linear, 1–1.3 mm long, pubescent, persistent; pedicels 4–6 mm, patent-pubescent. Posterior sepal keel-shaped, 3.5–4 mm long, patent-pubescent; wing sepals suborbicular to obovate-orbicular, 8–11 × 7–9 mm, conspicuously and densely reticulate, patent-pubescent; anterior sepals ovate-lanceolate, 3.5 mm long, patent-pubescent, free. Upper petals obliquely oblanceolate, 5–5.5 × 2 mm, pubescent from the waist downwards; carina 6.5–7 × 2.5 mm; crest 1–2 mm long, fimbriate. Stamens 8. Capsule ovate-elliptic in outline, 5.5–6 × 4.5–5 mm, asymmetrically bilobate, very narrowly winged (wing ± 0.1 mm wide), densely patent-pubescent. Seeds ovoid-ellipsoid, 4–4.5 × 0.9–2.2 mm, white-sericeous; caruncle asymmetric, with three appendages, usually two longer, sometimes almost up to as long as the length of the seed and the other one shorter (0.5–2 mm long).

UGANDA. Karamoja District: Lumeno R., Sept. 1958, *Wilson* 628! & Karasuk, Amudat, Apr. 1961, *Tweedie* 2124!
KENYA. Naivasha District: Suswa Plains, Suswa–Magadi, Oct. 1938, *Bally* 8431!; Masai District: Esakut Hill, Magadi road, 23 May 1964, *Archer* 438!; Kwale District: Mbuguni area, 3 Jan. 1993, *Luke* 3494!
TANZANIA. Tanga District: Mnyusi, Sept. 1954, *Sensei* 1834!; Kilosa District: Ilonga, 2 June 1967, *Robertson* 672!; Lindi District: Tendaguru, *Henning & Janeusch* 323!
DISTR. **U** 1; **K** 1–4, 6, 7; **T** 3, 6, 8; not known elsewhere
HAB. *Acacia-Commiphora* bushland, wooded and bushed grassland, woodland, scattered tree grassland, secondary bushland; 0–1500 m

SYN. *Polygala wadibomica* Chodat in Mém. Soc. Phys. Hist. Nat. Genève 31 (2): 320, t. 27/11 (1893). Type: Tanzania, Wadiboma (?=Kwediboma), *Fischer* s.n. (B†, holo.)
    P. hennigii Chodat in E.J. 48(1–2): 312 (1912). Type: Tanzania, Lindi District: Tendaguru, *Henning & Janeusch* 323 (B†, holo.)

NOTE. The type [Tanzania, Lushoto District: Kwa Mshusa, Handei, *Holst* s.n. (B†, holo.)] of *P. usambarensis* Gürke [in Abh. Preuss. Akad. Wiss.: 64 (1894); in P.O.A. C: 283 (1895); T.T.C.L.: 457 (1949); Paiva in Fontqueria 50: 312 (1998)] has disappeared and there are no duplicates. It is most likely *P. kilimandjarica* Chodat, which is very polymorphic and common in that area.

13. **Polygala senensis** *Klotzsch* in Peters, Reise Mossamb. Bot. 1: 113 (1861); Exell in F.Z. 1, 1: 315, t. 56/10 (1960); Thulin, Fl. Somalia 1: 82 (1993); U.K.W.F. ed. 2: 56 (1994); Paiva in Fontqueria 50: 177, t. 30/e-f (1998), *pro parte* excl. syn. *P. obtusissima.* Type: Mozambique, Sena, *Peters* s.n. (B†, holo.)

Perennial, occasionally annual, herb up to 60 cm tall, often with annual shoots produced from a woody rhizome. Stems slender, branched, densely crisped-pubescent, sometimes patent-pubescent. Leaves alternate, subsessile (petiole up to 0.1 mm, pubescent), narrowly oblong-lanceolate to oblanceolate, 10–30 × 5–10 mm, rounded to obtuse, base cuneate, densely pubescent on both sides. Flowers pinkish to yellowish, in lateral racemes 1.5–8 cm long, rachis pubescent; bracts ovate-acuminate, 1.2–1.5 mm long, pubescent, persistent; bracteoles ovate-acuminate, 1–1.3 mm long, pubescent, persistent; pedicels 4–6 mm, pubescent. Posterior sepal keel-shaped, 3.5–4 mm × 2.5 mm, pubescent; wing sepals suborbicular to obovate-orbicular, 8–10 × 7–8 mm, conspicuously and densely reticulate, pubescent; anterior sepals keel-shaped, 3.5 × 2.5 mm, pubescent, free. Upper petals obliquely oblanceolate, 5.5–6 × 2–2.5 mm; carina 6–6.5 mm × 2.5 mm; crest 2–3 mm long, fimbriate. Stamens 8. Capsule broadly oblong-elliptic in outline, 6–7.5 × 4.5–5.5 mm, emarginate, very narrowly winged (wing 0.1 mm wide), densely pubescent. Seeds ovoid-ellipsoid, 4–4.5 × 1.5–2 mm, white-sericeous; caruncle asymmetric, with three appendages, two humped at the apex of the seed and extending almost to the base of the seed, the other one short too (0.5–1.5 mm long). Fig. 5: 2, p. 11.

KENYA. Northern Frontier District: 46 km from Garissa on Hagadera road, 29 May 1977, *Gillett* 21218!; Kitui District: Endau Hill, 19 Nov. 1979, *Gatheri, Mungai & Kibui* 79/60!; Tana River District: South side of Galole, 22 Dec. 1964, *Gillett* 16535!
TANZANIA. Rufiji District: Rufiji, 20 Feb. 1931, *Musk* 26!; Kilwa District: Kingupura Forest, 1 Mar. 1976, *Vollesen* 3306! & 10 May 1972, *Ludanga* 1385!
DISTR. **K** 1, 4, 7; **T** 6, 8; Somalia, Mozambique, Zimbabwe and South Africa
HAB. Scattered tree grassland, grassland, often on sandy soils; 0–500 m

SYN. *Polygala multiflora* Mattei in Boll. Reale Orto Bot. Giardino Colon. Palermo 7(4): 177 (1908). Type: Somalia, Goscia, between Giumbo and Torda, *Macaluso* 116 (PAL!, holo.; K!, photo.), *non* Poir. (1804)
   P. matteiana Pampani in Boll. Soc. Bot. Ital. 1915(1–2): 16 (1915). Type: Somalia, Goscia, between Giumbo and Torda, *Macaluso* 116 (PAL!, holo.; K! photo.)
   P. rogersiana Bak.f. in J.B. 56: 5 (1918). Type: Mozambique, Vila Machado, *Rogers* 4505 (BM!, holo.; K!, iso.)

NOTE. The type [Tanzania, Zanzibar, *Hildebrandt* 905 (B†, holo.)] of *P. gagnebiniana* Chodat [Chodat in Mém. Soc. Phys. Hist. Nat. Genève 31 (2): 321, t. 27/12 (1893); T.T.C.L.: 456 (1949); U.O.P.Z.: 419 (1949); Paiva in Fontqueria 50: 308 (1998)] has disappeared. I saw material collected by *Hildebrandt* in the Island of Zanzibar with the same number 905 (BM, BREM, K, W). This material is *P. sphenoptera* Fresen. and does not agree with Chodat's original description of *P. gagnebiniana*, even with the drawing of the seed he presents in his monograph [in Mém. Soc. Phys. Hist. Nat. Genève 31 (2): t. 27/12 (1893)]. Chodat's original description of *P. gagnebiniana* and this drawing agree with *Polygala senensis* Klotzsch, which does not occur on Zanzibar. There are two explanations for this: either Chodat has cited a wrong specimen or *Hildebrandt* 905 (B†) was not the same species as the duplicates [*Hildebrandt* 905 (BM, BREM, K, W], which I have seen.

14. **Polygala sphenoptera** *Fresen.* in Mus. Senckenb. 2: 274 (1837); Petit in F.C.B. 7: 251, t. 14 (1958); Exell in F.Z. 1, 1: 315, t. 56/13 (1960); F.P.U.: 52 (1962); Blundell, Wild Fl. E Africa t. 719 (1978); U.K.W.F. ed. 2: 55 (1994); Thulin, Fl. Somalia 1: 82 (1993); Paiva in Fontqueria 50: 182, t. 34/a (1998); Gilbert in Fl. Eth. 2, 1: 180, fig. 24.2/1–4 (2000). Type: Ethiopia, between Halei and Tembien, *Rüppell* s.n. (FR!, holo.)

Annual or perennial herb or shrublet, 15–60(–100) cm tall. Stems ± erect, slender, branched, crisped-pubescent to patent-pubescent, rarely nearly glabrous. Leaves alternate, shortly petiolate, linear-lanceolate, narrowly oblong-elliptic to elliptic, 15–60(–80) × 3–8(–13) mm, subacute to rounded, base cuneate, crisped-pubescent to patent-pubescent on both sides; petiole 0.5–0.8 mm, pubescent. Flowers pink to purple, in lateral racemes 2–10(–18) cm long, ± densely flowered, rachis crisped-pubescent to patent-pubescent; bracts ovate-acuminate, 1.5–2 mm long, usually pubescent, ciliolate, persistent, bracteoles ovate, 1–1.3 mm long, usually pubescent, ciliolate, persistent; pedicels 4–6(–8) mm, usually glabrous, but often pubescent. Posterior sepal keel-shaped, 3–3.5 × 1.5–2 mm, pubescent or glabrescent; wing sepals suborbicular, 4–8 × 7–7.5 mm, shortly unguiculate, frequently with purple margin when dried, inconspicuously and laxly reticulate, glabrous or pubescent at the base; anterior sepals keel-shaped, 3–3.3 × 1.5–2 mm, pubescent or glabrescent, free. Upper petals obliquely oblanceolate, 3.5–4 × 1.5–2 mm, puberulous towards the base; carina 4.5–5 × 2.5–3 mm; crest 2–3 mm long, fimbriate. Stamens 8. Capsule broadly oblong-elliptic in outline, 4–6 × 3–4 mm, asymmetrically emarginate, narrowly winged (wing 0.2–0.3 mm wide), sometimes ciliolate. Seeds ovoid-ellipsoid, 2–3.5 × 1.5 mm, white-sericeous; caruncle subsymmetric, with three appendages, two longer, 0.5–1 mm long, the other shorter or almost absent. Fig. 5: 3, p. 11.

UGANDA. Karamoja District: Moroto, Kasimeri Estate, May 1971, *Wilson* 2041!; Bunyoro District: Bulisa, Jan. 1941, *Purseglove* 1107!; Mengo District: Kyadondo, Bogolobi, 6 July 1969, *Rwaburindore* 56!
KENYA. West Suk District: Marich Pass, 26 Oct. 1977, *Carter & Stannard* 48!; Naivasha District: E of Lake Naivasha, near Marina, 18 Apr. 1968, *Mwangangi* 714!; Kwale District: Cha Shimba Forest, 1 Feb. 1953, *Drummond & Hemsley* 1074!
TANZANIA. Mwanza District: Mabale, Mbarika, 21 Apr. 1953, *Tanner* 1411!; Kondoa District: Sambala, 25 Mar. 1929, *Burtt* 2127!; Mbeya District: near Igawa, 18 Jan. 1957, *Richards* 7914!; Zanzibar: Kidichi, 14 July 1961, *Faulkner* 2873!
DISTR. U 1, 2, 4; K 1–7; T 1–7; Z; from Ethiopia and Sudan south to South Africa
HAB. Evergreen bushland, wooded grassland, grassland; 0–2600 m

SYN. *Polygala quartiniana* Quart.-Dill. & A.Rich. in Ann. Sci. Nat. Bot., Sér. 2, Bot. 14: 263 (1840); T.T.C.L.: 456 (1949). Type: Ethiopia, Tigré, Assai, *Quartin-Dillon* s.n. (P!, holo.)
 P. quartiniana Quart.-Dill. & A.Rich. var. *minor* Chodat in Mém. Soc. Phys. Hist. Nat. Genève 31 (2): 328 (1893). Type: Ethiopia, Saganeïti, *Schweinfurth & Riva* 964 (B†, syn.; K! isosyn.), 1389 (B†, syn.)
 P. ukambica Chodat in Mém. Soc. Phys. Hist. Nat. Genève 31 (2): 329, t. 27/21 (1893). Type: Kenya, Kitui District: Ukamba, *Hildebrandt* 2785 (G!, syn.); South Africa, Natal, Weenen, *Wood* 4433 (G!, syn.)
 P. tristis Chodat in Bull. Herb. Boiss. 4(12): 903 (1896), *non* Chodat (1900). Type: Mozambique, Boroma, *Menyhart* 811 (Z!, holo.)
 P. ehlersii sensu De Wild. in Ann. Mus. Congo, Sér. 4, 2: 99 (1913), *non* Gürke (1894)
 P. fischeri Gürke in E.J. 70: 185 (1932); Paiva in Fontqueria 50: 183 (1998) Type: Tanzania, Tanga District: Pangani, *Stuhlmann* 455 (B†, holo.; K!, neo.)
 P. sphenoptera Fresen. var. *minor* (Chodat) Chiov. in Miss. Biol. Borana: 97 (1939)
 P. sphenoptera Fresen. var. *fischeri* (Gürke) E.M.A.Petit in F.C.B. 7: 252 (1958)

NOTE. The type [Tanzania, Mpororo, *Stuhmann* 2107 (B†, holo.)] of *P. stuhlmannii* Gürke [in P.O.A. C: 234 (1895); T.T.C.L.: 457 (1949); Paiva in Fontqueria 50: 312 (1998)] has disappeared and there are no duplicates. I found material collected by *Schlieben* 5116 (B, BM, M) determined as *P. stuhlmannii* Gürke, which is *P. sadebeckiana* Gürke. But the description in the protologue, mainly the one of the leaves, does not agree with *P. sadebeckiana* Gürke. It is, perhaps, *P. sphenoptera* Fresen., which is a very polymorphic species.
 The same happened with the type [Kenya, Kilimanjaro, *Volkens* 330 (B†, holo.)] of *P. lentiana* Gürke [in P.O.A. C: 234 (1895); Paiva in Fontqueria 50: 309 (1998)], which is *P. sphenoptera* Fresen. or *P. ehlersii* Gürke, but probably the former.
 I also did not find the type [Tanzania, Amani, *Warnecke* 409 (B†?, holo.)] of *P. ruderalis* Chodat [in Engl. E.J. 48(1–2): 315 (1912); Paiva in Fontqueria 50: 312 (1998)], which is almost certainly *P. sphenoptera* Fresen.

15. **Polygala ehlersii** *Gürke* in E.J. 19, Beibl. 47: 36 (1894); Paiva in Fontqueria 50: 183 (1998). Type: Tanzania, Kilimanjaro, *Ehlers* 68 (B†, holo.)

Annual or perennial herb, scrambling and semi-scandent, up to 15 cm tall. Stems slender, branched, crisped-pubescent. Leaves alternate, petiolate, ovate-elliptic to ovate-lanceolate, 10–40 × 4–15 mm, acuminate, base cuneate or subobtuse, crisped-pubescent on both sides; petiole 1–1.5 mm, crisped-pubescent. Flowers pink, in lateral racemes 2–10 cm long, flexuous, lax (space between two flowers up to 3 cm) and few-flowered, with 2–9 flowers and the rachis crisped-pubescent; bracts ovate-acuminate, 1.5–2 mm long, pubescent, persistent, bracteoles ovate, 1–1.3 mm long, pubescent, persistent; pedicels 4–10 mm, pubescent, but often pubescent. Posterior sepal keel-shaped, 3–3.5 × 1.5–2 mm, sparsely pubescent; wing sepals suborbicular, 5.5–7 × 5–6.5 mm, inconspicuously and laxly reticulate, glabrous; anterior sepals keel-shaped, 2.5–3 × 1.5–2 mm, sparsely pubescent, free. Upper petals obliquely oblanceolate, 4–4.5 × 1.5–2 mm, puberulous towards the base; carina 4.5–5 × 2.5–3 mm; crest 2–3 mm long, fimbriate. Stamens 8. Capsule broadly elliptic to almost orbicular in outline, 5–5.5 × 4.5–5 mm, asymmetrically emarginate, narrowly winged (wing 0.5–0.8 mm wide), ciliolate. Seeds ovoid-ellipsoid, 3–3.5 × 1.5 mm, shortly sericeous-pubescent; caruncle subsymmetric, with three appendages, two long, 0.5–1 mm long, the other one shorter or almost absent.

KENYA. Kericho District: Sambret Catchment, SW Mau Forest, 2 Aug. 1962, *Kerfoot* 3931!; Masai District: Loitokitok ridge, 29 Nov.. 1932, *Rogers* 107!; Teita District, Mt Vuria, 9 Feb. 1966, *Gillett, Burtt & Osborn* 17081!
TANZANIA. Mbulu District: Ufiome Mt, 12 Oct. 1930, *Burtt* 2371!; Lushoto District: Usambara Mts, near Magamba Peak, 24 Apr. 1968, *Renvoize & Abdallah* 1755!
DISTR. **K** 3–7; **T** 2, 3; not known elsewhere
HAB. Dry open places in short grassland, forest clearings and forest margins; 1800–2700 m

SYN. *Polygala kaessneri* Gürke in Bigger, Checklist Fl. Kilimj.: 8 (1968). Type: Kenya, Masai, Loitokitok, *Rogers* 107 (K!, holo.)

16. **Polygala vittata** *Paiva* in Fontqueria 50: 183, t. 6/t; t. 34/c; t. 35 (1998). Type: Kenya, Nakuru District: Eastern Mau Reserve, *Maas Geesteranus* 6162 (K!, holo.; LD!, iso.)

Perennial, prostrate and rosette-forming herb. Stems slender, annual shoots produced from a long woody rootstock, up to 30 cm long, crisped-pubescent. Leaves alternate, shortly petiolate (petiole 0.5–1 mm, crisped-pubescent), very narrowly elliptic to linear, 10–25 × 0.5–3.5 mm, slightly apiculate and mucronate, base cuneate, sparsely crisped-pubescent, revolute, sometimes entirely concealing the lower surface. Flowers purple, purplish-violet or pink, in slender lateral racemes 2–5.5 cm long, rachis crisped-pubescent; bracts lanceolate-linear, 1.5 mm long, persistent, bracteoles linear, 0.7–1 mm long, persistent; pedicels 2–4 mm, crisped-pubescent. Posterior sepal keel-shape, 2–3 × 1.5–1.8 mm, crisped-pubescent; wing sepals obliquely obovate-orbicular, 4–5 × 3.5–4 mm, with two red or purple-brown stripes down the centre which delimit a middle fringe, inconspicuously and laxly reticulate, glabrous; anterior sepals elliptic, 2–3 mm long, crisped-pubescent on the back, free. Upper petals obliquely lanceolate, 3.7–4 × 2 mm, broader near the apex, with a semi-circular incision at one side of the waist, pubescent from the waist downwards; carina 4.5 × 2 mm; crest 2–3 mm long, fimbriate. Stamens 8. Capsule elliptic in outline, 4.5–5 × 3 mm, ± symmetric and emarginate, narrowly winged (wing 0.2–0.5 mm wide), ciliate. Seeds ellipsoid, 2.5–3 × 1.5 mm, white-sericeous; caruncle asymmetric, with three short appendages, 0.4–0.5 mm long. Fig. 5: 4, p. 11 & Fig. 14, p. 34.

UGANDA. Karamoja District: Kadam, 1936, *Eggeling* 2735!; Ankole District: Isingiro, May 1939, *Purseglove* 694!; Masaka District: Kabula, Sept. 1945, *Purseglove* 1822!
KENYA. Turkana District: Murua Nysigar Peak, 24 Sept. 1963, *Paulo* 1024!; Nakuru District: Endabarra, Mau Forest, 22 Jan. 1946, *Bally* 4920!; Masai District: Narok–Olokurto road, km 19, 17 May 1961, *Glover, Gwynne & Samuel* 1215!
DISTR. **U** 1–4; **K** 1–6; Rwanda
HAB. Grassland, especially in rocky sites; 1150–3350 m

17. **Polygala arenaria** *Willd.*, Sp. Pl., ed. 5(5), 3: 380 (1802); Petit in F.C.B. 7: 248 (1958); Exell in F.Z. 1, 1: 320, t. 57/17, t. 58/C (1960); U.K.W.F. ed. 2: 55 (1994); Paiva in Fontqueria 50: 186, t. 34/d (1998); Gilbert in Fl. Eth. 2, 1: 184, fig. 24.3/5–6 (2000). Type: Guinea, sine loc., *sine coll.* (B!, holo.)

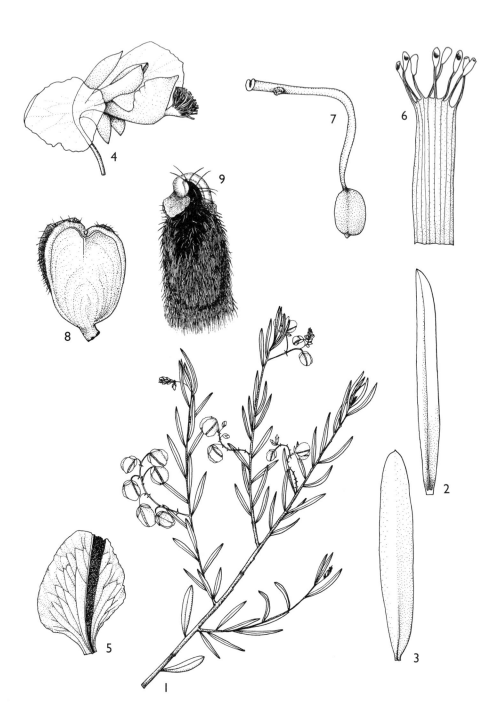

FIG. 14. *Polygala vittata* – **1**, habit × 2 ; **2**, leaf × 6; **3**, leaf × 6; **4**, flower × 12; **5**, wing sepal × 12; **6**, stamens × 25; **7**, pistil × 25; **8**, capsule × 25; **9**, seed × 25. All from *Geesteranus* 6162. Drawn by M. Lameiras.

Annual herb, 5–40 cm tall. Stems slender, of sympodial growth, pseudo-di- or trichotomously branched, apparently terminating in an inflorescence and growth continuing by lateral branches, usually exceeding the primary one, crisped-pubescent to nearly patent-pubescent. Leaves alternate, petiolate (petiole 1–3 mm, crisped-pubescent), linear-elliptic to oblanceolate, 10–60(–80) × 2–20 mm, subacute to rounded, base cuneate, densely crisped-pubescent to nearly glabrous on both sides. Flowers white, dull orange or lilac to pale purple, in dense terminal racemes, often globose-ovoid, 1.5–3 cm long, peduncle 3–10 mm, crisped-pubescent, rachis crisped-pubescent; bracts lanceolate, 1.5–2 mm long, usually pubescent, persistent, bracteoles lanceolate, 1–1.3 mm long, usually pubescent, persistent; pedicels 2–3(–5) mm, usually crisped-pubescent. Posterior sepal keel-shaped, 2–2.5(–3) × 1 mm, pubescent or glabrescent; wing sepals suborbicular to obliquely elliptic, (3.5–)5–7 × (2.5–)3–5(–5.5) mm, shortly stipitate, somewhat pubescent; anterior sepals keel-shaped, 1.2–2 × 1–1.3 mm, free. Upper petals obliquely elliptic, constricted below the middle, (2–)2.5–3 × 1–1.5 mm; carina 2.5–3 × (1–)1.5–2 mm; crest 1–2 mm long, fimbriate. Stamens 8. Capsule broadly obovate-elliptic in outline, (2.5–)3–4 × 2–3 mm, narrowly winged (wing 0.2–0.3 mm wide), ciliate. Seeds ovoid-ellipsoid, 2–2.5 × 1–1.3 mm, white-sericeous; caruncle subsymmetric, 0.5–0.8 mm long, with 3 small appendages. Fig. 5: 5, p. 11.

UGANDA. Karamoja District: Labwor, 4 May 1940, *A.S. Thomas* 3707!; Teso District: Soroti, 3 Sept. 1958, *Symes* 406!; Mengo District: 2 km N of Kakoge, 22 Dec. 1955, *Langdale Brown* 1813!
KENYA. West Suk District: Kongolai Escarp., July 1961, *Lucas* 218!; Kwale District: Shimba Hills, valley W Kidian, Aug. 1967, *Makin* 425! & outside Kaya Kinondo, 24 July 1993, *Luke* 3572
TANZANIA. Musoma District: Majita, Nyambono, 5 June 1959, *Tanner* 4327!; Kigoma District: Kitwe Point, 6 km of Kigoma, 9 Apr. 1994, *Bidgood & Vollesen* 3090!; Ulanga District: Machipi, 11 Sept. 1959, *Haerdi* 336/0!
DISTR. **U** 1–4; **K** 2, 7; **T** 1, 3, 4, 6, 8; widespread throughout tropical Africa, from Mauritania to Ethiopia and S to Angola and Mozambique
HAB. Grassland, scattered tree grassland, ruderal sites; 0–1700 m
USES. The ash of this species is used in Tanzania against puff-adder bite.

18. **Polygala welwitschii** *Chodat* in Mém. Soc. Phys. Hist. Nat. Genève 31 (2): 341, t. 27/46 (1893); Paiva in Fontqueria 50: 186, t. 34/e (1998). Type: Angola, Pungo Andongo, *Welwitsch* 1007 (B†, holo.; BM!, lecto.; LD!, LISU!, M!, isolecto.)

Annual erect herb, 6–18 cm tall. Stems slender, usually pseudo-di- or trichotomously branched, apparently terminating in an inflorescence and growth continuing by lateral branches, usually exceeding the primary one, crisped-pubescent. Leaves alternate, petiolate (petiole 1–3 mm, crisped-pubescent), lanceolate to oblong-elliptic, 15–30(–60) × 1–10 mm, subacute to rounded, base cuneate, glabrous or very minutely pubescent on both sides. Flowers whitish, pinkish or lilac, in much congested terminal racemes, often subglobose, 8–25 mm long, rachis crisped-pubescent; bracts lanceolate, 1–1.5 mm long, usually pubescent, persistent; bracteoles lanceolate, 1 mm long, usually pubescent, persistent; pedicels 1.5–3 mm, usually crisped-pubescent. Posterior sepal keel-shaped, 1.5–2 × 0.8 mm, pubescent or glabrescent; wing sepals elliptic to suborbicular, 3–4 × 2–3 mm, pubescent towards the base; anterior sepals keel-shaped, 1–1.5 mm long, free. Upper petals obliquely elliptic, constricted below the middle, 2.5 × 1.5–2 mm, pubescent at the base; carina 2.5 × 1.25 mm; crest 1.5–2 mm long, fimbriate. Stamens 8. Capsule elliptic to suborbicular in outline, 2–2.5 × 2–2.8 mm, glabrous, narrowly winged (wing 0.2 mm wide). Seeds ellipsoid to subglobose, 1.5–2.3 × 0.7–1 mm, brownish silky-pubescent, rarely puberulous; caruncle subsymmetric, 0.4–0.6 mm long, with three small appendages, 0.2–0.3 mm long.

subsp. **pygmaea** (*Gürke*) *Paiva* in Fontqueria 50: 189, t. 34/g (1998). Type: Tanzania, Bukoba District: Bukoba, *Stuhlmann* 3763 (B†, holo.; K!, lecto.; BM!, isolecto.)

Usually a dwarf herb, 3–9 cm tall. Stems simple or branched from the base, sometimes pseudo-di- or trichotomously branched. Leaves 15–60 mm long, longer than the proximal inflorescence; wing sepals 2–3 × 2–2.8 mm; seeds (1.5–)1.6–2.2 × 0.6–0.9, whitish-silky-pubescent, rarely puberulous. Fig. 5: 6, p. 11.

UGANDA. Masaka District: NW Lake Nabugabo, 9 Oct. 1953, *Drummond & Hemsley* 4708!
TANZANIA. Bukoba District: Mhutwe, June 1931, *Haarer* 2026!; Mpanda District: Mpanda, 18 Mar. 1958, *Jones* 56!
DISTR. **U** 4; **T** 1, 4; widespread throughout tropical Africa south of the Equator
HAB. Wooded grassland and grassland; 1050–1500 m

SYN. *Polygala pygmaea* Gürke in P.O.A. C: 234 (1895); Petit in F.C.B. 7: 248 (1958); Exell in F.Z. 1, 1: 323, t. 57/18 (1960)

NOTE. Subsp. *welwitschii* is an Angolan endemic and an erect herb up to 18 cm tall, with stems usually pseudo-di- or trichotomously branched, wing sepals 3–4 × 2–3 mm, seeds brownish-silky-pubescent.

19. **Polygala melilotoides** *Chodat* in E.J. 48: 320 (1920); Petit in F.C.B. 7: 246, fig. 7A (1958); Exell in F.Z. 1, 1: 323, t. 57/19 (1960); Paiva in Fontqueria 50: 189, t. 17/b; t. 34/h-k (1998). Type: Congo-Kinshasa, Katanga, Senga West, *Kassner* 2938 (B†, holo.; BM!, lecto.; BR!, isolecto.)

Annual herb, 3–15 cm tall. Stems slender, usually pseudo-di- or trichotomously branched, apparently terminating in an inflorescence and growth continuing by lateral branches, usually exceeding the primary one, crisped-pubescent. Leaves alternate, petiolate (petiole 1–2 mm, crisped-pubescent), elliptic to broadly elliptic, 10–30(–40) × (3–)5–12(–20) mm, subacute to rounded, base cuneate, glabrous or very minutely pubescent on both sides. Flowers whitish, pinkish or lilac, in much congested terminal racemes, often subglobose, 7–20 mm long, rachis crisped-pubescent; bracts lanceolate, 0.7–1 mm long, usually pubescent, persistent; bracteoles lanceolate, 0.5 mm long, usually pubescent, persistent; pedicels 1.5–3 mm, usually crisped-pubescent. Posterior sepal keel-shaped, 1.5–2 × 0.5–1 mm, pubescent or glabrescent; wing sepals elliptic to suborbicular, (2–)2.5–3.5 × 2–3 mm, pubescent towards the base; anterior sepals keel-shaped, 1–1.5 mm long, free. Upper petals obliquely obovate, constricted below the middle, 1.5–2.2 × 0.7–1.5 mm, pubescent at the base; carina 2.5–3 × 1.2–1.5 mm; crest 0.7–1 mm long, fimbriate. Stamens 8. Capsule elliptic to ovate in outline, 1.5–2.5 × 1.5–2 mm, glabrous, narrowly winged (wing 0.2 mm wide). Seeds ovoid to subglobose, 1–1.8(–2) × 0.6–0.8(–1) mm, glabrous, rarely puberulous; caruncle subsymmetric, subglobose, 0.4–0.6 mm long, with the appendages hiding the caruncle, 0.4–0.5 mm long. Fig. 5: 7, p. 11.

UGANDA. Busoga District: Lake Victoria, Lolui I., 8 May 1964, *Jackson* s.n.!
TANZANIA. Ufipa District: 30 km on Tatanda–Sumbawanga road, 27 Apr. 1997,, *Bidgood et al.* 3504!; Njombe District: Ruhudji, Lupembe, May 1931, *Schlieben* 839!; Songea District: between Kwamponjore and Songea, 14 Mar. 1956, *Milne-Redhead & Taylor* 9207!
DISTR. **U** 3; **T** 4, 7, 8; tropical Africa, S of the Equator and North of the tropic of Capricorn
HAB. Wooded grassland and grassland; 800–1700 m

SYN. *Polygala melilotoides* Chodat var. *major* Bak.f. in J.B. 56: 5 (1918). Type: Congo-Kinshasha, Katanga, Lubumbashi, *Rogers* 10887 (BM!, holo.; K!, iso.)

NOTE. *Polygala welwitschii* with its two subspecies and *P. melilotoides* are very close, and not easy to distinguish. The seed is the most important organ to differentiate them.

20. **Polygala albida** *Schinz* in Verh. Bot. Vereins Province Brandenburg 29: 53 (1888); Exell in F.Z. 1, 1: 320, t. 57/21 (1960); U.K.W.F. ed. 2: 55 (1994); Paiva in Fontqueria 50: 190, t. 7/d; t. 12a; t. 36/a, b (1998); Gilbert in Fl. Eth. 2, 1: 183 (2000). Type: Namibia, Olukonda, *Schinz* 506 (G†, holo.; K!, lecto.)

Annual herb, 5–40(–60) cm tall. Stems slender, branched, the secondary flowering branchlets shorter than the terminal primary one, crisped-pubescent. Leaves alternate, petiolate (petiole 1–3 mm, crisped-pubescent), linear to narrowly elliptic, 15–80(–100) × 2–15 mm, subacute to rounded, base cuneate, glabrous or, sometimes minutely pubescent on the midrib and towards the base. Flowers white, greenish white, pale pink to pale purple, in dense terminal and lateral racemes 2–7 cm long, rachis crisped-pubescent; bracts lanceolate, 1.5–2 mm long, usually glabrous or laxly ciliate, persistent, bracteoles lanceolate, 1–1.5 mm long, usually glabrous or laxly ciliate, persistent; pedicels deflexed, 1.5–4(–5) mm, usually crisped-pubescent. Posterior sepal keel-shaped, 3.5–4 × 1 mm, pubescent or glabrescent; wing sepals obliquely ovate-elliptic, 6–8(–9) × 5–6 mm, with a prominent venation, usually a few hairy towards the base; anterior sepals keel-shaped, 2–3 × 1.2–1.5 mm, free. Upper petals obliquely elliptic, notched on one side, 2–2.2 × 1–1.5 mm; carina 3–4 × 2 mm; crest 2 mm long, fimbriate. Stamens 8. Style 3–4 mm long. Capsule broadly elliptic in outline, 4.5–5 × 4–4.5 mm, winged (wing 0.5–1 mm wide), ciliate. Seeds ellipsoid, 4–5 × 1.5–2.2 mm, densely hairy (hairs 1.5–2 mm long); caruncle subsymmetric, kidney-shaped, with three small appendages.

DISTR. Widespread throughout tropical Africa, from Ethiopia to Namibia and South Africa

Racemes usually dense; capsule 4.5–5 × 4–4.5 mm, with a marginal
  wing 0.5–1 mm wide; seeds 4–5 × 1.5–2.2 mm, with hairs
  1.5–2 mm long . . . .   a. subsp. *albida*
Racemes usually laxe; capsule 3.5–4 × 3.5 mm, with a marginal
  wing up to 0.5 mm wide; seeds 3–4 × 1.3–1.5 mm, with
  hairs 0.5–1 mm long . . . .   b. subsp. *stanleyana*

a. subsp. **albida**

Racemes usually dense. Posterior sepal 3.5–4 mm long; wing sepals 6–8(–9) × 5–6 mm; anterior sepals 2–3 mm long. Capsule broadly elliptic in outline, 4.5–5 × 4–4.5 mm, narrowly winged (wing 0.5–1 mm wide). Seeds elliptic in outline, 4–5 × 1.5–2.2 mm, with hairs 1.5–2 mm long. Fig. 6: 1, p. 13.

TANZANIA. Moshi District: Marangu, 12 Oct. 1981, *Archbold* 2874!; Ufipa District: Milepa, 20 Feb. 1950, *Bullock* 2508!; Mbeya District: Mbozi, 10 May 1975, *Hepper, Field & Mhoro* 5492!
DISTR. **T** 2, 4, 7–8; widespread throughout tropical Africa, from Ethiopia to Namibia and South Africa
HAB. Woodland, wooded grassland, cultivated ground and road-sides; 900–1800 m

SYN. *Polygala livingstoniana* Chodat in Mém. Soc. Phys. Hist. Nat. Genève 31(2): 339, t. 27/38 (1893). Type: Angola, Massumba, *Pogge* 27 (B†, syn.); Namibia, Amboland, Aukondie, *Schinz* 506 (G†, syn.); South Africa, Transvaal, Hoggefeld, *Rehmann* 6731 (G†, syn.)
  [*P. arenaria sensu* Hiern, Cat. Afr. Pl. Welwitsch 1(1): 43 (1896), *pro parte*, excl. specim. *Welwitsch* 1002, *non* Willd. (1802)]

b. subsp. **stanleyana** (*Chodat*) Paiva in Fontqueria 50: 191, tab. 7/e; t. 12/b; t. 36/ c, d (1998); Gilbert in Fl. Eth. 2, 1: 184, fig. 24.3/1–4 (2000). Type: Angola, Massumba, *Pogge* 28 (B†, syn.); Angola, Pungo Andongo, *Welwitsch* 1015 (B†, *pro parte*, syn.); Angola, Pungo Andongo, *Welwitsch* 1017 (B†, syn.; BM!, neo.; LISU!, isoneo.)

Racemes usually lax. Posterior sepal 2.5–3.5 mm long; wing sepals 4–5.5(–6) × 3.5–4.5 mm; anterior sepals 2–2.5 mm long. Capsule elliptic in outline, 3.5–4 × 3.5 mm, narrowly winged (wing up to 0.5 mm wide). Seeds elliptic in outline, 3.5–4 × 1.3–1.5 mm, with hairs 0.5–1 mm long. Fig. 6: 2, p. 13.

UGANDA. West Nile District: Maracha Rest Camp, 7 Aug. 1953, *Chancellor* 128!; Toro District: N Ruwenzori, Karangora peak, Aug. 1954, *Osmaston* 3956!; Mengo District: Kawanda, Mar. 1936, *Chandler* 1593!

KENYA. Uasin Gishu: Soy Pool, Nov. 1948, *Bickford* B6541!; Machakos District: Ithaba Plains, 19
Apr. 1938, *Bally* 7903!; Masai District: Chyulu South, 21 June 1938, *Bally* 7946!
TANZANIA. Lushoto District: Mlalo Mission, 29 Jan. 1987, *Kisena* 429!; Ufipa District: Ilema Gap,
Rukwa road, 12 Mar. 1959, *Richards* 11176!; Songea District: Matengo Highlands, 13 km N of
Miyau, 1 Mar. 1956, *Milne-Redhead & Taylor* 8778!
DISTR. **U** 1–4; **K** 2–6; **T** 1–8; tropical Africa, from Tropic of Cancer to Namibia and Mozambique
HAB. Woodland, grassland, cultivated ground and road-sides; 500–2300 m

SYN. *Polygala stanleyana* Chodat in Mém. Soc. Phys. Hist. Nat. Genève 31(2): 340, t. 27/39
(1893); Petit in F.C.B. 7: 250 (1958)
P. stanleyana Chodat var. *angustifolia* Chodat in Mém. Soc. Phys. Hist. Nat. Genève 31(2):
340 (1893), *nom. illeg.*
P. modesta Gürke in E.J. 19, Beibl. 47: 35 (1894), *non* Miq. (1845). Type: Tanzania,
Muengue Mt, near Kwa Ngowe [localities not traced], *Volkens* 340 (B†, holo.; K!, lecto.;
BM!, E!, isolecto.)
[*P. persicariifolia sensu* De Wild. in Ann. Mus. Congo, sér. 5, 1: 51 (1903), *pro parte, non*
DC. (1824)]

21. **Polygala schweinfurthii** *Chodat* in Mém. Soc. Phys. Hist. Nat. Genève 31(2): 338,
t. 27/37 (1893); Petit in F.C.B. 7: 249 (1958); Paiva in Fontqueria 50: 191, t. 36/e (1998).
Type: Sudan, Bongo, Addai, *Schweinfurth* 2527 (B†, holo.; K!, lecto.; E!, isolecto.)

Annual herb, 15–40(–60) cm tall. Stems slender, simple or branched, the
secondary flowering branchlets shorter than the terminal primary one, crisped-
pubescent. Leaves alternate, petiolate (petiole 3 mm, crisped-pubescent), linear-
lanceolate to oblong lanceolate, 20–100(–120) × 3–25 mm, subacute to rounded,
base cuneate, glabrescent to glabrous. Flowers white, mauve to purple, in lax
terminal and lateral racemes 2–10 cm long, rachis crisped-pubescent; bracts
lanceolate, 1.5–2 mm long, ciliate, persistent, bracteoles lanceolate, 0.5–1.5 mm
long, ciliate, persistent; pedicels 3–5 mm, usually crisped-pubescent. Posterior sepal
keel-shaped, 3–3.5 × 1.5–2 mm, glabrescent; wing sepals obliquely ovate, 6–9 × 5–6
mm, with ± inconspicuous venation, ± pubescent, at least at the base; anterior sepals
keel-shaped, 2.5–3 × 1.5–2 mm, free. Upper petals obliquely ovate, 2–3 × 1.5–2 mm;
carina 4.5–5 × 2 mm; crest 2–2.5 mm long, fimbriate. Stamens 8. Style 6–7 mm long.
Capsule obovate-elliptic in outline, 3.5–4 × 3.5 mm, winged (wing up to 0.5 mm
wide), ciliate. Seeds ellipsoid, 3.5–4 × 1.5 mm, densely hairy (hairs 0.2–0.8 mm long);
caruncle subsymmetric, kidney-shaped, with three small appendages. Fig. 6: 3, p. 13.

UGANDA. Acholi District: Ukidi Forest, Nov. 1862, *Speke & Grant* s.n.!; Teso District: Serere, Sept.
1932, *Chandler* 877!; Mengo District: 1.5 km E of Kakoge, 14 Dec. 1955, *Langdale-Brown* 1700!
DISTR. **U** 1, 3, 4; Cameroon, Central African Republic, Sudan, Congo-Kinshasha and Burundi
HAB. Wooded grassland; 600–1000 m

SYN. [*P. arenaria sensu* Oliv., F.T.A. 1: 128 (1868), *pro parte, quoad Grant & Speke* s.n., *non*
Willd. (1802)]

22. **Polygala persicariifolia** *DC.*, Prodr. 1: 326, 64 (1824) as *persicariaefolia*; Chodat
in Mém. Soc. Phys. Hist. Nat. Genève 31(2): 331, t. 27/22–23 (1893); Petit in F.C.B.
7: 254 (1958); Exell in F.Z. 1, 1: 322, t. 57/16 (1960); F.P.U.: 52 (1962); U.K.W.F. ed.
2: 55 (1994); Paiva in Fontqueria 50: 192, t. 36/g (1998); Gilbert in Fl. Eth. 2, 1: 184,
fig. 24.3/7–10 (2000). Type: Nepal, *Wallich* s.n. (G-DC!, holo.)

Annual herb, up to 125 cm tall. Stems slender, simple or ± branched, crisped-
pubescent to glabrous. Leaves alternate, petiolate (petiole 1–2 mm, crisped-
pubescent), lanceolate to oblong-lanceolate, (20–)25–80 × 5–30 mm, acuminate, base
cuneate, sparsely pubescent to glabrescent. Flowers greenish white, pink, mauve to
pale purple, in terminal and lateral racemes 3–12 cm long, rachis crisped-pubescent;
bracts lanceolate, 1.5–2 mm long, persistent, bracteoles lanceolate, 0.5–1 mm long,
persistent; pedicels 3–5 mm, usually crisped-pubescent. Posterior sepal keel-shaped,

3–3.5 × 1–1.5 mm, glabrescent or ciliate; wing sepals obliquely suborbicular-ovate to broadly elliptic, 5–8 × (3.5–)4–6 mm, with prominent venation, glabrous or ciliate; anterior sepals keel-shaped, 2.5–3 × 1.5–2 mm, glabrescent or ciliate, free. Upper petals obliquely and irregularly elliptic, 2.5–4 × 1.5 mm, notched on one side; carina 4–6 × 1.5 mm; crest 2–3 mm long, fimbriate. Stamens 8. Capsule broadly-elliptic to suborbicular in outline, 5–7 × 4–5.5 mm, winged, wing up to 0.5 mm wide, ciliate. Seeds ellipsoid, 3–3.5 × 1.5 mm, with white silky hairs 0.2 mm long; caruncle subsymmetric, kidney-shaped, with three small appendages. Fig. 6: 4, p. 13.

UGANDA. Karamoja District: between Akisim and Kangole, June 1950, *Eggeling* 5986!; Toro District: Bwamba, Bubundu, 23 Sept. 1932, *Thomas* 704!; Mbale District: Kapchorwa, 7 Sept. 1954, *Lind* 232!
KENYA. Trans Nzoia District: NE Elgon, 35 km from Kitale on Suam road, Aug. 1971, *Tweedie* 4096!; South Kavirondo District: Lambwe, July 1934, *Napier* 6703!
TANZANIA. Ngara District: Bugufi, Nyamyaga, Nyakiziba, 27 Apr. 1950, *Tanner* 4900!; Mpanda District: Mahali Mts, Kasangazi, 30 Sept. 1958, *Jefford, Juniper & Newbould* 2777!; Iringa District: Kimiramatonge Circuit, 9.5 km from Msembe, 6 Mar. 1970, *Greenway & Kanuri* 14044!
DISTR. **U** 1–4; **K** 3, 5; **T** 1, 4, 6–8; widespread throughout tropical Asia and tropical Africa
HAB. Wooded grassland, grassland; 750–2100 m

SYN. *Polygala figarina* Webb, Fragm. Fl. Aethiop.-Aegypt.: 31 (1854). Type: Ethiopia, Fazogl, *Figari* s.n. (FT!, holo.; P!, iso.)
    P. hypericoides Webb, Fragm. Fl. Aethiop.-Aegypt.: 31 (1854). Type: Sudan, Kordofan, *Schimper* s.n. (FT!, holo.)
    P. granulata A.Rich., Tent. Fl. Abyss. 1: 39 (1847). Type: Ethiopia, Scholoda Mt, *Schimper* 1225 (P!, holo.; BM!, E!, K!, M!, OXF!, W!, iso.)
    P. persicariifolia DC. var. *densiflora* Chodat in Mém. Soc. Phys. Hist. Nat. Genève 31(2): 333 (1893). Type: Malawi, Shire Mt, *Buchanan* s.n. (BM!, lecto.)
    P. persicariifolia DC. var. *granulata* (A.Rich.) Chodat in Mém. Soc. Phys. Hist. Nat. Genève 31(2): 333 (1893)
    P. persicariifolia DC. var. *granulata* (A.Rich.) Chodat forma *latifolia* Chodat in Mém. Soc. Phys. Hist. Nat. Genève 31(2): 333 (1893). Type: Ethiopia, Matama, *Schweinfurth* 926 (G, lecto.; BM!, BREM!, G!, K!, W!, isolecto.)
    P. punctulata Hochst. in Flora 29, Intell. 1(2): 30 (1841), *nom. inval.* Type: Ethiopia, Scholoda Mt, *Schimper* 20 (P!, holo.; K!, iso.)
    P. persicariifolia DC. var. *punctulata* Chodat in Mém. Soc. Phys. Hist. Nat. Genève 31(2): 333 (1893). Type: Ethiopia, Scholoda Mt, *Schimper* 20 (P!, holo.; K!, iso.)
    P. persicariifolia DC. var. *wallichiana* Chodat forma *fazogliana* Chodat in Mém. Soc. Phys. Hist. Nat. Genève 31(2): 333 (1893). Type: Ethiopia, *sine loc.*, *Figari* s.n. (P!, holo.; FT!, iso.)
    P. persicariifolia DC. var. *fazogliana* (Chodat) T.Durand & Schinz, Consp. Fl. Afric. 1(2): 242 (1898).
    P. persicariifolia DC. var. *granulata* Chodat forma *macrophylla* Lanza & Mattei in Boll. Reale Orto Bot. Giardino Colon. Palermo 8(1–3): 80 (1909). Type: Ethiopia, Amasen, Filfil, *Lanza* 148 (FT!, holo.)
    P. spilophylla Steud., Nom. Bot., ed. 2, 2: 373 (1824), *nom. nud.*

23. **Polygala kasikensis** *Exell* [in De Wild. & Staner, Contr. Fl. Katanga, suppl. 4: 46 (1932), *nom. nud.*] in J.B. 70: 168 (1932); Petit in F.C.B. 7: 261 (1958); Paiva in Fontqueria 50: 201, t. 37/f (1998). Type: Congo-Kinshasa, upper Katanga, *De Witte* 412 (BR!, holo.; BM!, iso.)

Annual herb, up to 100 cm tall. Stems slender, simple or ± branched, glabrescent to glabrous. Leaves alternate, shortly petiolate (petiole 0.5–1 mm), linear to linear-lanceolate, 100–200 × 3–10 mm, acute, base cuneate, glabrescent to glabrous. Flowers pale purple, in many-flowered, simple or branched, ± dense, terminal racemes, narrowly conical to cylindric, 20–25 cm long, rachis patent-pubescent; bracts and bracteoles similar, ovate-lanceolate, 6–10 × 2.5–3.5 mm, persistent; pedicels 5–15 mm, usually patent-pubescent. Posterior sepal keel-shaped, 6–6.5 × 4–4.5 mm; wing sepals suborbicular, 14–14.5 mm in diameter; anterior sepals keel-shaped, 5.5–6 × 3.5–4 mm, free. Upper petals obliquely elliptic, 9 × 4 mm, pubescent; carina 12 mm long; crest

4–6 mm long, fimbriate. Stamens 8. Capsule obovate in outline, 8–9 × 5.5–6 mm, winged (wing up to 0.5 mm wide), ciliate. Seeds cylindric, 4–5 × 2 mm, with white silky hairs, 0.5 mm long; caruncle subsymmetric, kidney-shaped, with three very short appendages. Fig. 6: 5, p. 13.

TANZANIA. Ufipa District: Mbisi Forest, near Sumbawanga, 17 June 1960, *Leach & Brunton* 10064! & Mbizi Mts, 29 Apr. 1997, *Bidgood et al.* 3578!; Mbeya District: Mbozi, *sine die, Horsbrugh-Porter* s.n.!
DISTR. **T** 4, 7; Congo-Kinshasa
HAB. Forest edge, soil pockets in rock, grassland; 1800–2200 m

24. **Polygala ruwenzoriensis** *Chodat* in J.B. 34: 199 (1896); T.T.C.L.: 455 (1949); Petit in F.C.B. 7: 261 (1958); F.P.U.: 52 (1962); Paiva in Fontqueria 50: 202, t. 37/g (1998). Type: Uganda, Toro District, Mubuku Valley, *Scott Elliot* 7543 (K!, holo.; BM!, iso.)

Shrublet, 0.5–3.5 m tall. Stems subwoody, ± branched, pubescent to subtomentose. Leaves alternate, shortly petiolate (petiole 0.5–1 mm), linear to oblong-lanceolate, 2–10 × 0.3–2 cm, acuminate to acute, base cuneate, pubescent to glabrescent. Flowers violet, mauve to purple-red, in many-flowered, simple or few-branched, ± dense, narrowly ellipsoid, up to 15 cm long terminal racemes, rachis patent-pubescent; bracts ovate-lanceolate to lanceolate, 7–12 × 0.5–1.5 mm, acuminate or caudate-acuminate, hirsute; bracteoles ovate-lanceolate, 3–4 × 0.5–1 mm, hirsute, persistent; pedicels 8–16 mm, hirsute, persistent. Posterior sepal keel-shaped, 6–6.5 × 4–4.5 mm, pubescent; wing sepals suborbicular, 11–13 × 10–12.5 mm, glabrous; anterior sepals keel-shaped, 5–6 × 3–4 mm; pubescent, free. Upper petals obliquely elliptic, 6–7 × 3–5 mm; carina 10.5–12 mm long; crest 4–6 mm long, fimbriate. Stamens 8. Capsule obovate in outline, 6–8 × 4–6 mm, brownish, winged (wing up to 0.5 mm wide). Seeds cylindric, 3.2–4.2 mm, soft-pubescent (hairs ± 0.5 mm long, never exceeding the base of the seed); caruncle subsymmetric, kidney-shaped, with three very short appendages. Fig. 6: 6, p. 13.

UGANDA. Kigezi District: Behungi, 1 Dec. 1930, *Burtt* 2945! & Bwindi forest, Luhizha, 10 May 1992, *Cunningham* 4009! & Kachwekano Farm, May 1949, *Purseglove* 2887!
TANZANIA. Bukoba District: 16 km W of Bukoba, 18 Sept. 1961, *Rose* 10033!; Morogoro District: Uluguru Mts, 27 Dec. 1938, *Vaughan* 2660!; Mbeya District: Poroto Mts, May 1934, *Hornby* 655!
DISTR. **U** 2; **T** 1, 4, 6 7; Congo-Kinshasa, Rwanda, Burundi
HAB. Montane woodland and wooded grassland; 1500–3000 m

25. **Polygala macrostigma** *Chodat* in E.J. 48(1–2): 319 (1912); Petit in F.C.B. 7: 260 (1958); Exell in F.Z. 1, 1: 319, t. 56/1 (1960); Paiva in Fontqueria 50: 202, t. 12/c, t. 38/a (1998). Type: Congo-Kinshasa, Katanga, Kundelungu, *Kassner* 2771 (B†, holo.; K!, lecto.; BM!, BR!, E!, P!, isolecto.)

Annual herb, 1–3 m tall. Stems erect, simple or few-branched, sparsely pilose, mainly on the young parts, or glabrous. Leaves alternate, shortly petiolate (petiole 0.5–1 mm), linear and grass-like to very narrowly elliptic, 3–25 × 0.3–1.8(–2) cm, acute, base cuneate, slightly rough with short hairs, mainly on the upper surface, to glabrescent. Flowers purple to whitish, in many-flowered simple or few-branched terminal racemes up to 60 cm long, rachis pilose; bracts ovate-lanceolate, 4–10 × 0.5–1.5 mm, caudate with a thread-like apex, persistent; bracteoles ovate-lanceolate, 2.5–3 × 0.5–1 mm, persistent; pedicels 5–14 mm, pilose. Posterior sepal keel-shaped, 4.5–6.5 × 3–4 mm, pilose; wing sepals obliquely suborbicular, (9–)10–13.5 mm in diameter, glabrous or minutely ciliate; anterior sepals keel-shaped, 4–6 × 3–4 mm, pilose, free. Upper petals obliquely elliptic, 5–7 × 3–4 mm; carina 8.5–11 mm long; crest 3–4 mm long, fimbriate. Stamens 8. Capsule oblong-obovate in outline, 8–11 × 5.5–7.5 mm, winged, wing 0.5–0.8 mm wide, ciliate. Seeds cylindric, 3.2–4(–6) × 1.8–2 mm, with white silky hairs ± 2 mm long, exceeding the base of the seed; caruncle subsymmetric, kidney-shaped, with three very short appendages.

UGANDA. West Nile District: Oraba, Nov. 1938, *Hazel* 706!; Toro District: Fort Portal, Bwamba road, 23 May 1959, *Lind* 2553! Mbale District: Bukedi, July 1926, *Maitland* 1185!
KENYA. North Kavirondo District: Mumias, Oct. 1898, *Whyte* s.n.!
TANZANIA. Mwanza District: Mbarika, Lubiri, 7 May 1952, *Tanner* 753!; Kigoma District: Livandabe (Lubalisi) Mt, 31 May 1997, *Bidgood et al.* 4218!; Iringa District: N part ol Gologolo Mts, 12 Nov. 1970, *Thulin & Mhoro* 931!
DISTR.U 1–3; **K** 5; **T** 1–4, 6–8; from Sudan to Angola and Mozambique
HAB. Open woodland and wooded grassland, sometimes in marshes; 600–2500 m

SYN. *P. gomesiana sensu* T.T.C.L.: 455 (1949), *non* Oliv. (1868)

NOTE. I did not find the type [Tanzania, Kiriba, *Scott-Elliot* 8287 (B†?, holo.)] of *P. elliottii* Chodat [in Engl. E.J. 34: 199 (1896); Paiva in Fontqueria 50: 308 (1998)], but I am almost certain that it is conspecific with *P. macrostigma* Chodat. If so, *P. elliottii* Chodat [in Engl. E.J. 34: 199 (1896)] has priority over *P. macrostigma* Chodat [in E.J. 48(1–2): 319 (1912)]

26. **Polygala exelliana** *Troupin* in B.J.B.B. 19: 206 (1948), as *exelleana*; Petit in F.C.B. 7: 262 (1958); Exell in F.Z. 1, 1: 320, t. 56/2 (1960); Paiva in Fontqueria 50: 203, t. 38/c (1998). Type: Congo-Kinshasa, upper Katanga, Kitendwe-Kasiki, *De Witte* 373 (BR!, holo.; BM!, iso.)

Shrublet or perennial herb, 1–4 m tall. Stems erect, ± branched, densely patent- or crisped-pubescent. Leaves alternate, shortly petiolate (petiole 0.5–1 mm), linear-elliptic to oblong-lanceolate, 2–7 × 0.4–2 cm, acute or rounded, base cuneate, densely pubescent. Flowers purple, red or whitish, in congested many-flowered terminal racemes 4–15 cm long, rachis densely pubescent; bracts lanceolate, 3–4 × 0.5–1 mm, subacute, persistent; bracteoles ovate, 2–2.5 × 0.5–0.8 mm, persistent; pedicels 9–13 mm, slender, pilose. Posterior sepal keel-shaped, 4–6.5 × 3–3.5 mm; wing sepals obliquely suborbicular, 9–11 × 8–10 mm, with a few hairs towards the base to pubescent; anterior sepals keel-shaped, 4–5 × 3–3.5 mm, free. Upper petals obliquely elliptic and laterally notched, 8–11 × 3–4 mm; carina 10–11 mm long; crest 3–4 mm long, fimbriate. Stamens 8. Capsule obovate-elliptic in outline, 6–9 × 4–5 mm, narrowly winged (wing 0.2 mm wide), brownish, shiny, ciliate. Seeds cylindric, 3.5–4.5 × 1.5 mm, white, silky, short-pubescent; caruncle subsymmetric, kidney-shaped, with three very short appendages. Fig. 7: 1, p. 14.

TANZANIA. Bukoba District: Kiamawa, Sept.-Oct. 1938, *Gillman* 466!; Ufipa District: Malonje Mt, 12 Nov. 1949, *Bullock* 1909!; Rungwe District: Ulambya, Mlule, 3 Apr. 1972, *Leedal* 1110!
DISTR. **T** 1, 4, 7; Congo-Kinshasa, Zambia and Malawi
HAB. Open woodland and wooded grassland; 900–2300 m

SYN. [*Polygala gomesiana sensu* De Wild. in Ann. Mus. Congo, sér. 4, 1: 205 (1903), *non* Oliv. (1868)]
[*P. ruwenzoriensis sensu* Exell in J.B. 70: 182 (1932), *pro parte, non* Chodat (1896)]

27. **Polygala bakeriana** *Chodat* in J.B. 34: 199 (1896); Petit in F.C.B. 7: 263, t. 28 (1958); Paiva in Fontqueria 50: 203, t. 38/d (1998). Type: Burundi, *Scott Elliot* 8252 (B†, holo.; BM!, lecto.; BR!, K!, isolecto.)

Shrublet or perennial herb, 20–60 cm tall. Stems simple or ± branched, pubescent. Leaves alternate, shortly petiolate (petiole 0.5–1 mm), linear to oblong-lanceolate, 1.5–7 × 0.2–1 cm, rounded to acute, base cuneate, ± revolute when dried, pubescent to glabrescent. Flowers purple, lilac to pinkish, brownish when dried, in many-flowered, terminal racemes, branched at the base, 3–10 cm long, rachis pubescent; bracts lanceolate, 1.5–2 mm long, subacute, persistent; bracteoles ovate, 1.2–1.8(–2) mm long, persistent; pedicels 5–7 mm, slender, pubescent. Posterior sepal keel-shaped, 3–4 × 1.2–1.5 mm; wing sepals obliquely suborbicular, 6.5–8 mm in diameter; anterior sepals keel-shaped, 3–4 × 2–2.5 mm, free. Upper petals obliquely elliptic, 4–5.5 × 1.5–2.5 mm; carina 7–8 mm long; crest 3–4 mm long, fimbriate. Stamens 8. Capsule obovate in outline, 4–6 × 3–4 mm, brownish when dried, very narrowly winged, wing 0.2 mm wide, brownish-dull, ciliate. Seeds cylindric, 3–3.5 × 1–1.3 mm, with white silky hairs, 0.5 mm long; caruncle subsymmetric, kidney-shaped, up to 0.8 mm long, with three scarcely developed appendages. Fig. 7: 2, p. 14.

UGANDA. Ankole District: Lutoto hill, June 1938, *Eggeling* 3662!; Toro District: Kalinzu Forest Reserve, 31 July 1960, *Paulo* 617!
TANZANIA. Ngara District: Bugufi, Kirushya, 6 Apr. 1960, *Tanner* 4833!; Mpanda District: Mpanda–Uvinza road, Uzondo Plateau, 29 May 2000, *Bidgood, Leliyo & Vollesen* 4534!; Ufipa District: Mbizi Forest Reserve, 20 Jan. 1987, *Ruffo & Kisena* 2646!
DISTR. **U** 2; **T** 1, 4; Congo-Kinshasa, Burundi and Rwanda
HAB. Open woodland and grassland; 1000–2200 m

28. **Polygala nyikensis** *Exell* in Bol. Soc. Brot., sér. 2, 31: 10 (1957) & in F.Z. 1, 1: 317 (1960); Paiva in Fontqueria 50: 204, t. 39/a (1998). Type: Malawi, Nyika Plateau, *Benson* 1392 (BM!, holo.)

Perennial, prostrate herb. Stems spreading from a woody rootstock, 5–20 cm long, branched, crisped-pubescent. Leaves alternate, shortly petiolate (petiole 0.5–1 mm), narrowly oblong-elliptic, 5–25 × 2–6 mm, obtuse to acute and mucronulate, base cuneate, ± pubescent when young but soon becoming glabrous. Flowers pale mauve or magenta, in somewhat congested terminal racemes 2–5 cm long; bracts lanceolate, 1.5–2 mm long, subacute, persistent; bracteoles lanceolate, 1.2–1.8(–2) mm long, persistent; pedicels 3–5 mm, slender. Posterior sepal keel-shaped, 3–3.5 × 1.2–1.5 mm; wing sepals obliquely elliptic, 6–7 × 4.5–5 mm, glabrous; anterior sepals keel-shaped, 2.5 × 1.5 mm, free. Upper petals broadly falcate, 4.5–5.5 × 1.5–2.5 mm; carina 6–6.5 mm long; crest 2.5 mm long, fimbriate. Stamens 8. Capsule suborbicular in outline, 5.5–6 × 4.5–5 mm, narrowly winged (wing 0.2 mm wide), slightly pubescent. Seeds ellipsoid, 3–3.5 × 1.2–1.5 mm, white sericeous; caruncle symmetric, kidney-shaped, with three short appendages 0.7 mm long.

TANZANIA. Mbeya District: Mbeya Peak Forest Reserve, 3 July 1962, *Mgaza* 521!
DISTR. **T** 7; Zambia and Malawi
HAB. Upland grassland; no altitude given (elsewhere 1300–2500 m)

29. **Polygala usafuensis** *Gürke* in E.J. 30: 337 (1900); Petit in F.C.B. 7: 257, fig. 7E (1958); Exell in F.Z. 1, 1: 324, t. 56/15 (1960); Paiva in Fontqueria 50: 206, t. 12/d, t. 39/d (1998). Type: Tanzania, Mbeya District: Usafwa, *Goetze* 1032 (B†, holo.; BM!, lecto.; E, P, isolecto.)

Annual herb, 1.5–2 m tall. Stems erect, robust, simple or few-branched, densely patent-pubescent, rarely with a few crisped hairs, to glabrescent. Leaves alternate, shortly petiolate (petiole 0.5 mm), lanceolate to very narrowly elliptic, 10–70 × 0.5–10 mm, acute, base cuneate, pubescent to glabrescent. Flowers orange-pink to salmon-pink, in many-flowered simple or few-branched terminal racemes up to 35 cm long, rachis densely pubescent; bracts lanceolate, 1–2 mm long, pubescent, persistent; bracteoles linear-lanceolate, 0.5–1 mm long, pubescent, persistent; pedicels 2–3 mm, pubescent. Posterior sepal keel-shaped, 3–5 × 1–2 mm; wing sepals broadly elliptic, 6–8 × 4–6 mm; anterior sepals keel-shaped, 2.5–3.5 × 1–2 mm, free. Upper petals obliquely obovate, 4–7 × 2–3 mm, with a finger-like lobe at the apex, pubescent on the back; carina 6–8 mm long, pubescent on the back; crest 2.5–4(–5) mm long, fimbriate. Stamens 8. Capsule oblong-obovate in outline, 5–6(–7) × 2.5–3(–3.5) mm, yellowish when dried, unwinged or very narrowed winged, ciliate. Seeds cylindric,

(3–)3.5–5 × 1–1.5 mm, with white silky hairs, 1.5 mm long; caruncle subsymmetric, kidney-shaped, with scarcely developed appendages. Fig. 7: 3, p. 14.

TANZANIA. Mpanda District: Kahoko, 22 July 1859, *Newbould & Harley* 4535!; Mbeya District: Mbeya, 11 May 1956, *Milne-Redhead & Taylor* 10045!; Songea District: Luhira R., 29 Apr. 1956, *Milne-Redhead & Taylor* 9854!
DISTR. **T** 4, 5, 7, 8; from Sudan and Congo-Kinshasa to Angola and Mozambique
HAB. Open woodland and wooded grassland, sometimes in marshes or seepage grassland; 600–2500 m

SYN. *Polygala verdickii* Gürke in Ann. Mus. Congo, sér. 4, 1: 205 (1903). Type: Congo-Kinshasa, Katanga, Lukafu, *Verdick* 556 (BR!, holo.)
　P. riparia Chodat in E.J. 48 (1–2): 317 (1912). Type: Congo-Kinshasa, Katanga, *Kassner* 2836 (B†, BM!, BR!, K!, syn.); *Kassner* 2862a (B†, BM!, BR!, K!, syn.)
　P. heliostigma Chodat in Bull. Soc. Bot. Genève, sér. 2, 5: 190 (1913). Type: Congo-Kinshasa, Bukama, *Bequaert* 155 (BR!, holo.)
　[*P. guerkei sensu* Exell in De Wild. & Staner, Contr. Fl. Katanga, suppl. 4: 46 (1932), *non* Chodat (1912)]
　P. tanganyikensis Troupin in B. J. B. B.: 208 (1949). Type: Congo-Kinshasa, M'Vua, near Lake Tanganyika, *van Meel* 1091 (BR!, holo.)
　[*P. sparsiflora sensu* Exell in J.B. 70: 182 (1932), *pro parte*, *non* Oliv. (1868)]

30. **Polygala sparsiflora** *Oliv.* in F.T.A. 1: 127 (1900), *pro parte, quoad* var. α; Paiva in Fontqueria 50: 206, tab. 39/e (1998). Type: Sierra Leone, *Morson* s.n. (K!, holo.; BM!, iso.)

Annual herb, 30–60(–100) cm tall. Stems erect, slender, simple or few-branched, scarcely crisped-pubescent, rarely patent-pubescent. Leaves alternate, shortly petiolate (petiole 0.5 mm), linear to linear-lanceolate, 10–30 × 0.5–1.5 mm, acute, base cuneate, slightly pubescent to glabrous. Flowers pink, magenta to yellow, in many-flowered or few-flowered, simple lax, flexuous, terminal racemes (none on the axils of the upper leaves) up to 35 cm long, rachis pubescent; bracts linear-lanceolate, 0.8–1 mm long, pubescent, persistent; bracteoles linear-lanceolate, 0.5 mm long, pubescent, persistent; pedicels 2.5–3 mm, pubescent. Posterior sepal keel-shaped, 2–2.5 × 1.3–1.5 mm, pubescent; wing sepals broadly elliptic, (4.5–)6–7 × 3–4.5 mm, densely pubescent; anterior sepals keel-shaped, 1.5–2 × 1 mm, free. Upper petals obliquely obovate, 3.5–4 × 1.3–1.5 mm; carina 4–5 × 2.5–3 mm long, glabrous; crest 2–2.5 mm long, fimbriate. Stamens 8. Capsule narrowly obovate in outline, 4–4.5 × 2–2.5 mm, slightly emarginate, unwinged or very narrowly winged, pubescent. Seeds cylindric, 3–3.5 × 1–1.3 mm, with white silky hairs, 1–1.3 mm long; caruncle symmetric, kidney-shaped, with scarcely developed appendages. Fig. 7: 4, p. 14.

var. **ukirensis** (*Gürke*) *Paiva* in Fontqueria 50: 207, t. 39/f (1998). Type: Tanzania, North Mara District: Ukira, *Fisher* 28 (B†, holo.; W!, lecto.; BM!, K!, isolecto.)

Racemes usually dense, some of them on the axils of the upper leaves; pedicels 1.5–2 mm, a little longer than the bracts (1–1.5 mm); wing sepals 4.5–5.5 × 3–3.5 mm; carina 4–5 × 2.5–3 mm, rarely ciliate; crest 0.5–1.5 mm.

UGANDA. Lango District: Oruma, Moroto, Sept. 1935, *Eggeling* 2219!; Teso District, Soroti, 17 Sept. 1954, *Lind* 388!; Masaka District: Sango Bay, between Sisal Estate and Lake Victoria, 17 Aug. 1951, *Norman* 39!
KENYA. Nandi District: Nandi Forest, Aug. 1933, *Dale* 3182!; South Kavirondo District: Kavirondo, sine die, *Scott Elliot* 7150! & Trans Nzoia District: Maboonde near Kitale, Aug. 1955, *Tweedie* 1339!
TANZANIA. Bukoba District: Kagera, Minziro Forest Reserve, Kinwa Kyaishemweru forest area SE Minziro, 12 July 2001, *Festo, Bayona & Wilbard* 1593!; Kigoma District: 58 km S Uvinza, 31 Aug. 1950, *Bullock* 3280!; Mbeya District: between Mbozi and Lake Rukwa, 10 May 1975, *Hepper, Field & Mhoro* 5478!
DISTR. **U** 1, 3, 4; **K** 3, 5; **T** 1, 4, 7, 8; Tropical Africa N of Tropic of Capricorn

HAB. *Brachystegia* woodland, bushland and upland grassland, sometimes in temporary marshes and seepage areas; (275–)1000–2250 m

SYN. *Polygala ukirensis* Gürke in E.J. 14: 310 (1891); Petit in F.C.B. 7: 256 (1958); Exell in F.Z. 1, 1: 327, t. 56/12 (1960); U.K.W.F. ed. 2: 55 (1994)
  [*P. sparsiflora sensu* Exell in J.B. 70: 182 (1932), *pro parte, non* Oliv. (1868)]

NOTE. Var. *sparsiflora* is a West tropical African plant from low altitudes (up to 1000 m), and has lax racemes, pedicels 2.5–3 mm, wings 6–7 × 4–4.5 mm, carina 4.5–6 × 2–2.5 mm, crest 2–2.5 mm long; while var. *ukirensis* is widespread throughout tropical grassland of high altitudes (over 1000 m) and has dense racemes, pedicels 1.5–2 mm, wings 4.5–5.5 × 3–3.5 mm, carina 4–5 × 2.5–3 mm, crest 0.5–1.5 mm long. In West tropical Africa, it is sometimes, not easy to separate these two.
  Vollesen 4543 from Miombo Valley (T 6), in spite of growing in a seepage at low altitude, cannot belong to *P. usafuensis* Gürke. More material from this place is required.

31. **Polygala nambalensis** *Gürke* in Warb. Kunene-Samb.-Exped. Baum: 276 (1903); Petit in F.C.B. 7: 258, t. 7/F (1958); Exell in F.Z. 1, 1: 325, t. 56/11 (1960); Paiva in Fontqueria 50: 208 (1998). Type: Angola, Bié, R. Cuchi, *Baum* 871 (B†, holo.; BM!, lecto.; COI!, K!, isolecto.)

Annual herb, 40–200 cm tall. Stems slender, simple or few-branched, appressed-pubescent to glabrous. Leaves alternate, shortly petiolate (petiole 0.5 mm), linear to linear-lanceolate, (8–)20–80 × 0.3–2 mm, pointed, base cuneate, revolute, sparsely pubescent to glabrous. Flowers yellowish or pale blue, in many-flowered, simple or few-branched, one-sided terminal racemes up to 25 cm long, rachis densely patent-pubescent to glabrous; bracts linear-lanceolate, 1.7–2 mm long, persistent; bracteoles linear-lanceolate, 1.2–1.5 mm long, persistent; pedicels 1.5–2.5(–4) mm, pubescent. Posterior sepal keel-shaped, (2.5–)3–4 × (0.7–)1–1.5 mm, glabrous; wing sepals obliquely broadly elliptic, (5.5–)6–7 × 4–4.5 mm, glabrous, occasionally a few hairs towards the base; anterior sepals keel-shaped, 2–3 × 1–1.8 mm, glabrous, free. Upper petals obliquely ovate-lanceolate, 5–6 × 3.5–4 mm, notched at one side; carina 7–8 mm long; crest 3.5–5 mm long, fimbriate. Stamens 8. Capsule oblong-obovate in outline, 4–6 × (2–)2.5–3 mm, very narrowly winged, ciliate. Seeds cylindric, 2.7–4 × 1.25 mm, with white silky hairs, 1–1.3 mm long; caruncle conical, much longer than wide, 1–1.5 mm long, without membranous appendages.

TANZANIA. Ufipa District: Sumbawanga, 10 km on Tatanda–Mbala road, 24 Apr. 1997, *Bidgood et al.* 3423!
DISTR. **T** 4; Congo-Kinshasha, Angola, Zambia
HAB. *Brachystegia* woodland; ± 1800 m

SYN. *P. britteniana* Chodat [as "*brittoniana*"] var. *phyllostigma* Chodat in J.B. 48(1–2): 319 (1912). Type: Congo-Kinshasha, Upper Katanga, R. Kasanga, *Kassner* 2662 (B†, holo.; BM!, E!, K!, P!, iso.)
  [*P. sparsiflora sensu* Exell in J.B. 70: 182 (1932), *pro parte, non* Oliv. (1868)]

32. **Polygala ohlendorfiana** *Ecklon & Zeyher*, Enum. Pl. Afr. Austr. 1: 22 (1834); Exell in F.Z. 1, 1: 317 (1960); U.K.W.F. ed. 2: 56 (1994); Paiva in Fontqueria 50: 209, t. 39/1 (1998). Type: South Africa, Orange Mts, Winterberg, *Ecklon & Zeyher* s.n. (M!, lecto.; BREM!, CGE!, E!, K!, LD!, OXF!, W!, isolecto.)

Perennial, prostrate herb. Stems sprawling and spreading from a woody rootstock, 5–20 cm long, branched, crisped-pubescent and sparsely pilose. Leaves alternate, shortly petiolate (petiole 0.5–1.5 mm, pubescent), ovate, ovate-lanceolate to suborbicular, 5–18 × 3–14 mm, obtuse, rounded to subacute and mucronulate, rounded or obtuse to very slightly cordate, pilose or crisped-pubescent, mainly on the margins and midrib. Flowers purple, dark magenta-red or pink, in few-flowered terminal racemes 1.5–5 cm long; bracts lanceolate-linear,

1–1.5 mm long, sparsely pubescent, persistent; bracteoles lanceolate-linear, 0.7–1 mm long, sparsely pubescent, persistent; pedicels 3–5 mm, slender, pubescent. Posterior sepal keel-shaped, 2.5–2.7 × 1.2 mm, sparsely pubescent; wing sepals obliquely broadly-elliptic, 5–7 × 3.5–4.5 mm, glabrous or sparsely pubescent on the midrib and often ciliate; anterior sepals keel-shaped, 2–2.2 × 1 mm, pubescent, free. Upper petals obovate-spathulate, 3.5–3.8 × 1.5–2 mm; carina 4–4.5 mm long; crest 2.7 mm long, fimbriate, crest 1.5 mm long. Stamens 8. Capsule suborbicular in outline, 4–4.5 × 4 mm, narrowly winged (wing 0.2–0.5 mm wide), slightly pubescent and ciliate. Seeds broadly ovoid, 2–2.8 × 1.5–1.8 mm, white sericeous-pubescent; caruncle strongly asymmetric, comma-shaped, with three appendages 0.8–1 mm long. Fig. 7: 5, p. 14.

KENYA. Trans-Nzoia District: Kitale–Elgon, Nov. 1940, *Jex-Blake* B1259!
TANZANIA. Mbeya District: Kikondo, 18 Oct. 1956, *Richards* 6639!
DISTR. **K** 3; **T** 7; Zambia, Malawi, Mozambique, Zimbabwe and South Africa
HAB. Upland grassland, burnt grassland; 1800–2750 m

NOTE. The material from Kenya [*Jex-Blake* B1259 (K)] consists of a very bad specimen, but looks like *P. ohlendorfiana*, which extends the distribution of this species.

33. **Polygala stenopetala** *Klotzsch* in Peters, Reise Mossamb., Bot. 1: 114, t. 23 (1865) [as "*stenophylla*"]; Exell in F.Z. 1, 1: 331, t. 57/1 (1960); Blundell, Wild Fl. E Africa t. 558 (1978); Paiva in Fontqueria 50: 219, t. 41/b, c (1998). Type: Mozambique, Inhambane, *Peters* s.n. (B†, holo.; P!, lecto.)

Annual or perennial herb, up to 150 cm tall. Stems erect, slender, simple or few-branched, narrowly winged, glabrous. Leaves alternate, grasslike, shortly petiolate (petiole 0.5 mm), lanceolate to narrowly lanceolate, 5–30 × 2–7 mm, pointed, base cuneate, glabrous. Flowers blue to greenish-blue, in terminal simple or few-branched racemes up to 20 cm long; rachis glabrous; bracts linear-lanceolate, 1–2 mm long, caducous; bracteoles linear-lanceolate, 0.8–1 mm long, caducous; pedicels 3–4 mm, pubescent. Posterior sepal keel-shaped, 2–2.5 × (0.7–)1–1.5 mm, glabrous; wing sepals obliquely broadly elliptic, 5–8 × 4–5.5 mm, with purple-brown veining, glabrous; anterior sepals keel-shaped, 1.8–2 × 1–1.5 mm, glabrous, connate. Upper petals obliquely obovate, 3–3.5 × 2.5 mm; carina 6.5–7 × 3.5–4 mm; crest 2–2.5 mm long, fimbriate. Stamens 8. Capsule oblong-elliptic in outline, 5–6.5 × 3.5–4.5 mm, unwinged, glabrous. Seeds ellipsoid, 3–4 × 1.5 mm, with short brownish appressed hairs; caruncle asymmetric kidney-shaped, 0.6–0.7 mm long, almost without membranaceous appendages.

subsp. **stenopetala**

Leaves lanceolate to narrowly lanceolate, 5–30 × 2–7 mm; wing sepals blue to greenish-blue, 5–8 × 4–5.5 mm; upper petals 3–3.5 × 2.5 mm; carina 6.5–7 × 3.5–4 mm; crest 2–2.5 mm long; capsules 5–6.5 × 3.5–4.5 mm; seeds 3–4 × 1.5 mm, with rather short hairs. Fig. 8: 1, p. 15.

KENYA. Kwale District: Shimba Hills forest, 14 Jan. 1964, *Verdcourt* 3911! and Shimba Hills NR, Longomwagandi Forest, 22 Mar. 1968, *Magogo & Glover* 394
TANZANIA. Tanga District: Moa, 4 Aug. 1963, *Drummond & Hemsley* 3629!; Morogoro District: Morningside road, 25 Oct. 1934, *Bruce* 37!; Songea District: Matengo hills, Mtama, E Ndengo, on Songea road, 13 Jan. 1956, *Milne-Redhead & Taylor* 8243!
DISTR. **K** 7; **T** 3, 6–8; Zambia, Malawi, Mozambique, Zimbabwe, South Africa
HAB. *Brachystegia-Julbernadia* woodland, grassland and cultivated ground; 0–1900 m

SYN. [*P. rarifolia sensu* Chodat in Mém. Soc. Phys. Hist. Nat. Genève 31(2): 367 (1893), *pro parte*, quoad syn.]
[*Polygala stenophylla* Klotzsch, *non* DC. (1824)]
Poplygala viminalis Gürke in P.O.A. C: 234 (1895). Type: Tanzania, Amboni, *Holst* 2929a (B†, holo.; W!, lecto.)

P. viminalis Gürke var. *brachyptera* R.E.Fries in Wiss. Ergebn. Schwed. Rhod.-Kongo-Exped. 1: 114 (1914). Type: Zambia, Abercorn, *Fries* 1257 (UPS!, holo.)

NOTE. The Angolan endemic subsp. *casuarina* (Chodat) Paiva in Fontqueria 50: 219, t. 12/f, tab. 41/d (1998) [*P. viminalis* Gürke var. *casuarina* Chodat in E.J. 48(1–2): 322 (1912)] has the leaves linear, 35–45 × 1 mm; wing sepals often yellow, 7.5–8.5 × 5–6 mm; upper petals 5–6 × 2–2.5 mm; carina 8.5–10 × 4.5 mm; crest 3–4 mm long; capsules 6.5–7.5 × 3.5–4.5 mm; seeds 4.5–5 × 2–2.5 mm, with rather long hairs.

34. **Polygala multifurcata** *Mildbr.* in N.B.G.B. 12: 708 (1935); Paiva in Fontqueria 50: 224 (1998). Type: Tanzania, Iringa District: Ngololo, Lepembe, *Schlieben* 701 (B†, holo.; BM!, lecto., M!, isolecto.)

Perennial herb. Stems 20–35 cm long, simple or few-branched, grooved, 4-angled, sometimes narrowly winged when young, glabrous, shortly and sparsely pubescent when young. Leaves alternate, shortly petiolate, oblong-elliptic, oblong-obovate or obovate, 10–20 × 5–10 mm, rounded to obtuse, subcuneate, glabrous. Flowers pink, in lateral lax few-flowered racemes up to 10–15 cm long; rachis flexuous, winged, glabrous; bracts linear-lanceolate, 1–2 mm long, caducous; bracteoles linear-lanceolate, 0.8–1 mm long, caducous; pedicels 3–4 mm. Posterior sepal keel-shaped, 3 × 1–1.5 mm, glabrous; wing sepals broadly elliptic, 5 × 3.5 mm, greenish with green veins, glabrous; anterior sepals keel-shaped, 2 × 1–1.5 mm, glabrous, connate. Upper petals obliquely obovate, 4–4.5 × 4 mm; carina 4 × 2 mm; crest 1 mm long, fimbriate. Stamens 8. Capsule oblong-elliptic in outline, 4 mm 3–3.5 mm, bilobed, narrowly winged, glabrous. Seeds subcylindric, 3 × 1–1.5 mm, sericeous-pubescent; caruncle subglobose, 1 mm long, almost without membranous appendages.

TANZANIA. Iringa District: Dabaga Highlands, Udzungwa Scarp Forest Reserve, 30 Jan. 1971, *Mabberley* 632!; Udzungwa Mountains NP, Luhomero camp 134 to 137, 5 Oct. 2000, *Luke et al.* 7013; Kilosa District, Rubeho Mts, Ukwiva Forest Reserve, 31 May 2005, *Luke et al.* 11041
DISTR. **T** 7; not known elsewhere
HAB. Weed in plantations surrounded by rain forest, montane forest with grassy glades and swamps; 1800–2200 m

35. **Polygala petitiana** *A.Rich.*, Tent. Fl. Abyss. 1: 37 (1847); Petit in F.C.B. 7: 278, t. 29 (1958) ; Exell in F.Z. 1, 1: 334 (1960); F.P.U.: 52 (1962); U.K.W.F. ed. 2: 55 (1994); Paiva in Fontqueria 50: 226, t. 43/i (1998); Gilbert in Fl. Eth. 2, 1: 187, fig. 24.4/12–14 (2000). Type: Ethiopia, near Gafta, *Schimper* 1188 (P!, lecto.; BM!, FR!, K!, M!, OXF!, isolecto.)

Annual herb, up to 100(–150) cm tall. Stems erect, slender, simple or few-branched, glabrescent to glabrous. Leaves alternate, subsessile, linear to linear-lanceolate, (10–)20–40(–60) × 1–4.5 mm, pointed, base cuneate, glabrous. Flowers blue, sometimes white to yellowish, in terminal, simple or few-branched racemes up to 20(–35) cm long; rachis glabrous; bracts linear to linear-lanceolate, 2 mm long, caducous; bracteoles linear, 0.8–1 mm long, caducous; pedicels 1.5–4 mm, glabrous. Posterior sepal keel-shaped, (1.5–)2–3 × 1–2 mm, glabrous; wing sepals obovate-elliptic, (2–)3–5 × (0.8–)1.8–3 mm, with (3–)4–5 blue veins, usually not from the base, glabrous; anterior sepals keel-shaped, (1–)1.5–2 × 1.2–1.8 mm, glabrous, connate. Upper petals obovate-spathulate, 3–5 × (1–)1.7–3 mm; carina (2–)3–4.5(–5) × 2.5–4 mm; crest absent, exceptionally with a vestigial one. Stamens 8. Capsule oblong-elliptic in outline, 3–4 × (2–)2.2–2.5(–3) mm, emarginate, very narrowly winged, glabrous. Seeds ellipsoid, (2–)2.5–3(–3.5) × 1–1.5 mm, with long, white, silky hairs; caruncle asymmetrically kidney-shaped, 0.7–0.8 mm long, hairy, almost without membranous appendages.

P. petitiana is a variable species, but it is possible to divide the complex into two subspecies; the typical one widespread throughout tropical Africa and subsp. *parviflora*, an East African taxon, with few-flowered racemes and smaller flowers than the former.

a. subsp. **petitiana**

Stems not slender, ± thick; leaves linear-lanceolate, 1–4.5 mm wide; racemes many-flowered; wing sepals 3–5 × 1.8–3 mm, 4–5-veined usually not from the base; carina 3–4.5(–5) mm long; capsule 3–4 × 2.2–2.5(–3) mm; seeds 2.5–3 × 1–1.5 mm with the hairs clearly exceeding the base of the seed.

subsp. **petitiana** var. **petitiana**; Fig. 9: 1, p. 17.

Stems simple or few-branched in the upper part; racemes of the branches shorter than the one of the main stem; wing sepals 4–5 × 2–3 mm; capsule up to 3.5–4 × 2.5(–3) mm; seeds 3 × 1–1.5 mm, caruncle 0.6–0.8 mm long.

UGANDA. West Nile District: NW of Maracha Rest Camp, 3 Aug. 1953, *Chancellor* 102!; Kigezi District: Ruzhumbura, Nyakagyeme, Nov. 1946, *Purseglove* 2228!; Teso District: Serere, May 1932, *Chandler* 559!
KENYA. Trans Nzoia District: Kitale, near Crampton's Inn, 14 Sept. 1958, *Napper* 800!; Masai District: Loitokitok, NE Kilimanjaro, 27 Feb. 1933, *Rogers* 564!; Kwale District: Matuga, 31 July 1982, *Robertson* 3313!
TANZANIA. Ufipa District: Nsanga Mt, Malonje Plateau, 13 Mar. 1959, *Richards* 12138!; Dodoma District: Kazikazi, 20 May 1932, *Burtt* 3590!; Songea District: 17 km W Songea, 24 Feb. 1956, *Milne-Redhead & Taylor* 8742!; Zanzibar: Zanzibar I., 1931, *Vaughan* 1902!
DISTR. **U** 1–4; **K** 3–7; **T** 1–8; **Z**; widespread throughout tropical Africa
HAB. *Brachystegia-Julbernardia* woodland, grassland and cultivated ground; 0–2100 m

SYN. *Polygala tetrasepala* Webb, Fragm. Fl. Aethiop.: 33 (1854). Type: Ethiopia, near Gafta, *Schimper* 1188 (BM!, FR!, K!, M!, OXF!, P!, syn.); *Schimper* 1192 (P!, W!, syn.); Ethiopia, near Djeladjeranne, *Schimper* 1650 (K!, M!, P!, W!, syn.)
    P. volkensii Gürke in P.O.A. C: 234 (1895). Type: Tanzania, Kassodjo (= Kasoge?), *Stuhlmann* 2250 (B†, syn.), *Stuhlmann* 3435 (B†, syn.); Tanzania, Pangani District: R. Pangani, *Volkens* 332 (B†, syn.), *Volkens* 569 (B†, syn.)

subsp. **petitiana** var. **abercornensis** *Paiva* in Fontqueria 50: 228, t. 43/j (1998). Type: Zambia, Mbala [Abercorn], near Kambole Falls, *Richards* 8305a (K!, holo.). Fig. 9: 2, p. 17.

Stems branched nearly from the base; racemes of the basal branches longer than the one of the main stem; wing sepals 3.5 × 1.8 mm; capsule up to 3.5 × 2.3 mm; seeds 2.5–2.7 × 1 mm, caruncle 0.5–0.6 mm long.

TANZANIA. Ufipa District: Namwele, 24 Feb. 1950, *Bullock* 2573!
DISTR. **T** 4; Zambia
HAB. Grassland and wet places; 1500–1900 m

b. subsp. **parviflora** (*Exell*) *Paiva* in Fontqueria 50: 228, t. 43/k (1998). Type: Malawi, lower Kasupse, *Exell, Mendonça & Wild* 819 (BM!, holo.; LISC!, SRGH!, iso.). Fig. 9: 3, p. 17.

Stems very slender; leaves linear, 0.5–1 mm wide; racemes few-flowered, up to 10(–20) flowers; wing sepals 2–3 × 0.8–1.5 mm, 3-veined from the base; carina 2–2.5 mm long; capsule up to 3 2 mm; seeds 2–2.5 × 1 mm, with the hairs slightly exceeding the base of the seed.

UGANDA. Mengo District: Nambigirwa, Mar. 1923, *Maitland* 600!
KENYA. Uasin Gishu District: Uasin Gishu, Aug. 1931, *Harvey* 1378!
TANZANIA. Njombe District: Lumbila, 12 Aug. 1958, *Gilli* 297!; Songea District; Matagoro hills, 27 Feb. 1956, *Milne-Redhead & Taylor* 8752!
DISTR. **U** 4; **K** 3; **T** 7, 8; Sudan, Congo-Kinshasa, Burundi, Zambia, Malawi, Mozambique and Zimbabwe

HAB. *Brachystegia-Julbernadia* woodland and grassland; 750–2400 m

SYN. *P. petitiana* A.Rich. var. *parviflora* Exell in Bol. Soc. Brot., sér. 2, 31: 12 (1957) & in F.Z. 1,
      1: 334, t. 57/5 (1960). Type: Malawi, lower Kasupse, *Exell, Mendonça & Wild* 819 (BM!,
      holo.; LISC!, SRGH!, iso.)
   [*P. liniflora sensu* Exell in J.B. 64: 302 (1926)]

36. **Polygala xanthina** *Chodat* in E. J. 48 (1–2): 325 (1912); Petit in F.C.B. 7: 276
(1958); Exell in F.Z. 1, 1: 333 (1960); Paiva in Fontqueria 50: 229, tab. 43/1
(1998). Type: Congo-Kinshasa, Katanga, R. Kasanga, *Kassner* 2661 (B†, holo.; BM!,
E!, K!, lecto.)

Annual herb, up to 50(–70) cm tall. Stems arcuate-ascending, ridged, usually
branched from the base, glabrous. Leaves alternate, subsessile, linear-lanceolate to
very narrowly elliptic, (15–)20–80 × (2–)4–15 mm, acute, base cuneate, scabrid on
the margin, otherwise glabrous. Flowers bluish to greenish-purple, in terminal,
simple or few-branched racemes up to 15(–20) cm long; rachis glabrous; bracts
linear to linear-lanceolate, 1.2–2 mm long, caducous; bracteoles linear, 0.5–0.8 mm
long, caducous; pedicels 1.5–3 mm, glabrous. Posterior sepal keel-shaped, 2–3 ×
1.5–2 mm, glabrous; wing sepals ovate-elliptic, (3.5–)4.5–6 × 2.5–3 mm, with 3–5
greenish veins, glabrous; anterior sepals keel-shaped, 1.5–2 × 1.5–1.8 mm, glabrous,
connate. Upper petals obliquely obovate, 5–6.5 × 3–4 mm; carina 5–6 × 3–4.5 mm;
crest absent. Stamens 8. Capsule oblong-elliptic in outline, 4.5–5 × 3.5–4 mm,
narrowly winged, glabrous. Seeds ellipsoid, 3–3.5 × 1.2–1.5 mm, with white, silky
hairs; caruncle asymmetric kidney-shaped, 0.7–0.8 mm long, hairy, appendages
scarcely developed. Fig. 9: 4, p. 17.

UGANDA. Kigezi District: Kichwamba, Mar. 1946, *Purseglove* 1969!; Mengo District: Kijude, July
   1915, *Dümmer* 2604!
TANZANIA. Mpanda District: Mahali Mts, Utahya, 6 Sept. 1958, *Jefford, Juniper & Newbould* 2388!;
   Mbeya District: 14 on Tunduma-Sumbawanga road, 21 Apr. 1997, *Bidgood et al.* 3350!; Kilwa
   District: Machinga, 17 June 1953, *Anderson* 923!
DISTR. **U** 2, 4; **T** 4, 7, 8; Congo-Kinshasa, Burundi, Angola, Zambia, Malawi, Mozambique and
   South Africa
HAB. Woodland and grassland; 1000–1800 m

37. **Polygala gossweileri** *Exell* in J.B. 74: 133 (1936) & in F.Z. 1, 1: 335 (1960); Paiva
in Fontqueria 50: 234 (1998). Type: Angola, Bié, R. Colui, *Gossweiler* 2146 (BM!,
holo.; COI, iso.)

Perennial herb sending up annual shoots up to 15 cm long from a woody
rootstock and forming tufts. Stems erect or decumbent, slender, striate, glabrous.
Leaves alternate, subsessile, linear, 5–20 × 0.5–1 mm, acute, with a needle-like
point, base cuneate, margin revolute, glabrous. Flowers blue, solitary or in a lateral
simple few-flowered raceme, up to 1.5 cm long; rachis glabrous; bracts linear, 0.5
mm long, caducous; bracteoles linear, ± 0.2 mm long, caducous; pedicels 5–9 mm,
glabrous. Posterior sepal keel-shaped, 2.2 × 1.5 mm, ciliolate; wing sepals obliquely
elliptic, 4.5–5 × 2.5–3 mm, acute, 3–5-veined from the base, ciliolate; anterior
sepals keel-shaped, 1.7–2 × 1.5 mm, ciliolate, connate. Upper petals obliquely
obovate, 4–4.5 × 3–3.5 mm; carina 5–5.2 × 2.5–2.8 mm; crest 1.5 mm long, bilobed.
Stamens 6 fertile with 2 staminodes. Capsule oblong-elliptic in outline, 2.5–3 ×
2–2.5 mm, narrowly winged, glabrous. Seeds ellipsoid, 2–2.2 × 0.7–1 mm, with
white-sericeous hairs; caruncle asymmetrically kidney-shaped, 0.7 mm long, hairy,
appendages scarcely developed.

TANZANIA. Songea District: Ulamboni valley, 11 km W of Songea, 1 Jan. 1956, *Milne-Redhead &
   Taylor* 8018!

Distr. **T** 8; Angola, Zambia and Malawi
Hab. Grassland; ± 960 m

Syn. *Polygala huillensis sensu* Exell in J.B. 65: 347 (1927), *pro parte quoad specim. Gossweiler* 2146, *non* Oliv. (1868)

38. **Polygala luteo-viridis** *Chodat* in E.J. 48 (1–2): 323 (1912); Petit in F.C.B. 7: 273 (1958); Paiva in Fontqueria 50: 235, t. 44/g (1998). Type: Tanzania, Bukoba District: 34 km from Bukoba, *Haarer* 2151 (K!, neo.) & Bukoba, *Stuhlmann* 1011 (B†, syn.); Rwanda, Mohusi, *Wildbraed* 524 (B†, syn.)

Shrublet or perennial herb sending up annual shoots 5–25 cm long from a woody rootstock and forming tufts. Stems arcuate-ascending, slender, sparsely pubescent to glabrescent. Leaves alternate, petiolate (petiole 0.5–1 mm), oblanceolate to obovate-linear, exceptionally linear, 5–25 × (2–)3–6 mm, acute, base cuneate, margin revolute, sparsely pubescent to glabrescent. Flowers pink to yellow, in terminal or lateral simple or few-branched, many-flowered racemes up to 6 cm long; rachis sparsely pubescent to glabrescent; bracts linear, ± 0.7 mm long, caducous; bracteoles linear, 0.5 mm long, caducous; pedicels 2–3 mm, sparsely pubescent to glabrescent. Posterior sepal keel-shaped, 2–3 × 1.5–2 mm, glabrous; wing sepals ovate-elliptic, 4.5–5.5 × 2.5–3.5 mm, 3–5-veined from the base, sparsely pubescent to glabrescent; anterior sepals keel-shaped, 1.7–2 × 1.5–2 mm, glabrous, connate. Upper petals obliquely obovate, 4.5–6 × 3–3.5 mm; carina 5–6 × 2.7–3.5 mm; crest 1.5 mm long, bilobed. Stamens 6 fertile with 2 staminodes. Capsule oblong-elliptic to oblong-obovoid in outline, 3.5 × 2–2.5 mm, narrowly winged, glabrous. Seeds ellipsoid, 2.7 × 1.2–1.5 mm, with white, silky hairs; caruncle asymmetric kidney-shaped, ± 0.7 mm long, hairy, appendages scarcely developed. Fig. 9: 5, p. 17.

Uganda. Busoga District: Butembe Bunya, 16 Jan. 1953, *Wood* 591!; Masaka District: Lake Nabugabo, Aug. 1935, *Chandler* 1357!; Mengo District: Entebbe, Lake Victoria, 6 Nov. 1968, *Lye* 180!
Kenya. Kisumu-Londiani District: Yala R., on Kisumu–Kakamega road, May 1961, *Tweedie* 2148!; North Kavirondo District: Broderick Falls, May 1971, *Tweedie* 3953!
Tanzania. Mwanza District: near Mwanza, on Lake Victoria, 24 Oct. 1932, *Geilinger* 3268!; Iringa District: Imagi Mt, 48 km E of Iringa, 15 Dec. 1961, *Richards* 15687!
Distr. **U** 3, 4; **K** 5; **T** 1, 7; Rwanda
Hab. Grassland; 1100–1900 m

39. **Polygala acicularis** *Oliv.*, F.T.A. 1: 132 (1868); Petit in F.C.B. 7: 270 (1958); Paiva in Fontqueria 50: 236, t. 45/c (1998). Type: Uganda, West Nile District: near Madi, *Grant & Speke* 132 (K!, holo.)

Shrublet, rarely annual herb, 50–100 cm tall. Stems woody, erect, densely leafy, striate, glabrous. Leaves alternate, shortly petiolate (petiole 0.5–1 mm), linear to acicular, (10–)20–60 × 1–3 mm, mucronate, glabrous. Flowers whitish to blue-purplish, in dense terminal racemes 10–15(–30) cm long; rachis glabrous; bracts linear, 2–3 mm long, caducous; bracteoles linear, up to 1 mm long, caducous; pedicels 1.2–1.5 mm, glabrous. Posterior sepal elliptic, 3 mm long, minutely ciliate; wing sepals elliptic, 6–9 × 2.5–4 mm, glabrous, early caducous; anterior sepals elliptic, 2.5 mm long, minutely ciliate, connate. Upper petals obovate, 6.5–8 × 3.5–4 mm; carina 5–7.5 × 3–4 mm; crest 2.5–3 mm long, fimbriate. Stamens 6 fertile and glabrous with 2 staminodes pubescent. Capsule elliptic in outline, 5–7 × 4 mm, deeply notched, very narrowly winged, glabrous. Seeds ellipsoid, 4–6 × 2–2.5 mm, with white, silky indumentum; caruncle symmetric, 1.5–2 mm long; appendages absent. Fig. 10: 1, p. 18.

UGANDA. Acholi District: Gulu, 8 Dec. 1931, *Hancock* 2387!; Teso District: Serere, Jul.-Aug. 1932,
    Chandler 834!; Mengo District: Sungira, Nakasongola, 11 July 1956, *Langdale-Brown* 2201!
DISTR. U 1, 3, 4; Chad, Nigeria, Cameroon, Gabon, Central African Republic, Sudan, Congo-
    Kinshasa, Angola
HAB. Swampy grassland and bushland; 900–1200 m

40. **Polygala amboniensis** *Gürke* [in Abh. Königl. Akad. Wiss. Berlin: 22 (1894),
nom. nud.] in Engl., P.O.A. C: 234 (1895); Blundell, Wild Fl. E Africa t. 718 (1978);
Thulin in Fl. Somalia 1: 87 (1993); Paiva in Fontqueria 50: 237, t. 45/d (1998). Type:
Tanzania, Tanga District: Amboni, *Holst* 2548 (B†, holo.; BREM!, lecto.)

Perennial or annual herb, 30–70 cm tall. Stems erect, much spreading, branched,
striate, 4-angled, glabrous. Leaves alternate, shortly petiolate (petiole 0.5–1 mm),
linear to oblong-spathulate, (5–)10–30 × (0.5–)1–2(–3) mm, acuminate or emarginate
and mucronate, base cuneate, glabrous. Flowers bluish, pinkish mauve, salmon pink
or yellowish, in lax terminal racemes 12–30 cm long; rachis glabrous; bracts linear,
1–1.3 mm long, glabrous; bracteoles linear, 1 mm long, glabrous; pedicels 1.5–2 mm,
glabrous. Posterior sepal ovate, 1.5–1.8 mm long, ciliolate; wing sepals obovate-
elliptic, 3–3.5 × 1–1.3 mm, glabrous, early caducous; anterior sepals 1–1.2 mm long,
ciliolate, connate. Upper petals obovate-spathulate, 3–3.5 × 2 mm; carina 3.5–4 ×
1.8–2 mm; crest 1 mm long, fimbriate. Stamens 6 fertile with 2 staminodes, glabrous.
Capsule ovate in outline, 3.5–4(–5) × (2–)2.5–3 mm, deeply asymmetric-bilobate and
long acuminate (acumen up to 1–1.5 mm), glabrous, unwinged. Seeds conical, 2.5–3
× 1–1.2 mm, sericeous-pubescent; caruncle pyramidal, 0.8–1 mm long; appendages
0.4–0.6 mm long. Fig. 10: 2, p. 18.

KENYA. Northern Frontier District: Mandera, War Gedud, 1 May 1978, *Gilbert & Thulin* 1301!;
    Tana River District: Kurawa, 48 km S of Garsen, 5 Oct. 1961, *Polhill & Paulo* 594!; Kwale
    District, Marenji Forest Reserve, 23 May 2000, *Luke & Saidi* 6250
TANZANIA. Masai District: Longido Mt, above Longido Village, 31 Jan. 1969, *Richards* 23625!;
    Uzaramo District: Kunduchi, 19 May 1968, *Harris* 1770!; Kilwa District: Kilwa, Aug. 1873, *Kirk*
    s.n.!; Zanzibar: Marahubi, 22 Apr. 1961, *Faulkner* 2811!
DISTR. K 1, 7; T 2, 3, 6, 8; Z; Somalia
HAB. Bushland, secondary grassland, cultivated fields, on sandy soils and coastal dunes; 0–100
    (–850) m

SYN. *P. liniflora* Chodat in Mém. Soc. Phys. Hist. Nat. Genève 31 (2): 367 (1893), *pro parte*
    ["*linifolia*" in index]; U.K.W.F. ed. 2: 55 (1994)
    P. sennii Chiov., Fl. Somalia 2: 14/4 (1932). Type: Somalia, Jack Omisso, *Senni* 387*bis*
    (FT.!, holo.)

41. **Polygala conosperma** *Bojer* in Ann. Sc. Nat., sèr. 2; 4: 266 (1835); Paiva in
Fontqueria 50: 237, t. 45/e (1998). Type: Kenya, Mombasa, *Hildebrandt* 1925 (B†,
holo.; P!, lecto.; BREM!, K!, M!, W!, isolecto.)

Perennial or annual herb, 30–100 cm tall. Stems erect, simple or few-branched,
striate, glabrous. Leaves alternate, shortly petiolate (petiole 0.5–1 mm), linear, 25–60
× 0.5–1 mm, mucronate, base cuneate, margin revolute, glabrous. Flowers pale red
or yellowish, in lax terminal racemes 10–30 cm long; rachis glabrous; bracts linear-
lanceolate, 2–2.5 mm long, ciliate; bracteoles linear, 0.7–0.8 mm long, ciliate;
pedicels 1.5–2.5 mm, glabrous. Posterior sepal ovate, 2–2.5 mm long, ciliate; wing
sepals obovate-elliptic, 4–4.5 × 2.5 mm, 3–5-veined from the base, glabrous, persistent;
anterior sepals 1–1.5 mm long, ciliate, connate. Upper petals obovate-spathulate,
4.5–5 × 1.5–2 mm; carina 3.5–4 × 1.5–1.8 mm; crest 1 mm long, fimbriate. Stamens 6
fertile with 2 staminodes. Capsule obovate-elliptic in outline, 3.5 × 2–2.2 mm, slightly
asymmetric-bilobate, glabrous, very narrowly winged. Seeds obovoid-ellipsoid, 3–3.5
× 1.5 mm, with white silky hairs; caruncle almost symmetric, oblong kidney-shaped,
1 mm long; appendages 0.2 mm long. Fig. 10: 3, p. 18.

KENYA. Mombasa District: Likoni, July 1932, *Blake* 2278; Kwale District: Kaya Tiwi area, 30 Oct. 1992, *Luke* 3323!
TANZANIA. Pangani District: Bushiri, 11 Aug. 1950, *Faulkner* 624!; Zanzibar, 1847–1852, *Boivin* s.n. *pro parte*!
DISTR. **K** 7; **T** 3; **Z**; not known elsewhere
HAB. Edges of wooded grassland, secondary grassland not far from the coast; 0–10 m

SYN. *P. aphrodisiaca* Gürke [ex Pax in Engl, Pflanzenw. Ost-Afr. B: 514 (1895), *nom. nud.*] in P.O.A. C: 234 (1895): 367 (1893), ["*phrodisiaca*"]. Type: Tanzania, Pangani, *Stuhlmann* 98 (B†, holo.)
     P. liniflora Chodat in Chiov. Result. Sc. Miss. Stefanini-Paoli Somal. It. 1: 25 (1916)
     [*P. leptoclada sensu* Chodat in Bull. Herb. Boiss. 4(12): 908 (1896), *non* Bojer (1843)]

42. **Polygala muratii** *Jacq.-Fél.* in Bull. Soc. Bot. Fr. 99: 66/1 (1952); Paiva in Fontqueria 50: 239 (1998); Gilbert in Fl. Eth. 2, 1: 187, fig. 24.5/1–5 (2000). Type: Chad, Biltine plain, *Murat* 641 (P!, holo.)

Perennial or annual herb, 30–60 cm tall. Stems erect or ascending, much branched from near the base, striate, 4-angled, glabrous. Leaves alternate, subsessile, linear to oblong-spathulate, 10–35 × 1–3 mm, mucronate, base cuneate, margin revolute, glabrous. Flowers salmon pink, reddish-pink to yellow-orange, in lax terminal or lateral racemes 10–20 cm long; rachis glabrous; bracts linear, 1–1.3 mm long, glabrous, caducous; bracteoles linear, 0.7 mm long, glabrous, caducous; pedicels 2.5–3 mm, glabrous. Posterior sepal ovate, 1.5–1.8 mm long, ciliate; wing sepals obovate, 4.5–5.5 × 2.5–3 mm, with a long claw, 3-veined from the base and with anastomosing venation only near the base, glabrous, persistent; anterior sepals 1–1.3 mm long, ciliate, connate. Upper petals obovate-spathulate, 4.5–5 × 2.2–2.5 mm; carina 5–5.5 × 2.5 mm; crest 1.5 mm long, fimbriate. Stamens 6 fertile with 2 staminodes. Capsule ovate in outline, 3.5–4 × 2.5–3 mm, deeply asymmetric-bilobate and long acuminate (acumen up to 0.8 mm), glabrous, unwinged. Seeds conical, 2.5–3 × 1–1.3 mm, sericeous-pubescent, with the convex base completely covered by glands; caruncle pyramidal, up to 0.2 mm long, completely covered by hairs; appendages 0.2 mm long.

UGANDA. Karamoja District: Lodoketemit, Jan. 1964, *Napper* 1707! & Moroto, Kasimeri estate, May 1971, *J. Wilson* 2043!
KENYA. Northern Frontier District: Dandu, 5 May 1952, *Gillett* 13050!; Naivasha District: Suswa crater-rim, 23 Mar. 1963, *Bally* 12641!; Teita District: Ndi Mt, 16 Apr. 1960, *Verdcourt & Polhill* 2711!
TANZANIA. Mwanza District: Ukerewe I., 11 Oct. 1930, *Conrads* 5500!; Masai District: Longido Mt, 14 Jan. 1936, *Greenway* 4359!; Kilosa District: Masenge camp, Mapilingo R., 11 June 1969, *Sanane* 811!
DISTR.**U** 1; **K** 1–4, 6, 7; **T** 1, 2, 5, 6; Chad, Sudan and Ethiopia
HAB. Grassland, seepage grassland, thin soil over rock; 200–2000 m

SYN. [*Polygala liniflora sensu* Brown & Massey, Fl. Sudan: 67 (1929), *non* Chodat (1893)]
     [*P. nilotica sensu* A.Chev. & Jacq.-Fél. in Chev., Fl. Afr. Occ. Fr.: 271 (1938), *non* Chodat (1893)]

NOTE. *P. muratii*, *P. conosperma* (*P. liniflora*) and *P. amboniensis* have been mixed up and were usually determined as *P. liniflora*. The capsule of *P. conosperma* is slightly asymmetric-bilobate, while the other two species have the capsule strongly asymmetric-bilobate; its caruncle is kidney-shaped and almost glabrous and the seed is obovoid-ellipsoid, while *P. muratii* and *P. amboniensis* have the caruncle pyramidal and completely covered by hairs and the seeds conical. The latter two species are distinct in their habitat, as *P. amboniensis* occurs in lowlands not too far from the coast [0–100(–850) m] and *P. muratii* at higher altitudes (200–2000 m); and by the wing sepals, which are early caducous in the former, but persistent in the latter. There is another species in the area of this Flora with conical seeds and pyramidal caruncle: *P. irregularis*, but this one has the anterior sepals free, while the other two have the anterior sepals connate.

43. **Polygala irregularis** *Boiss.*, Diagn. Pl. Orient., ser. 1, 1: 8 (1843); Thulin, Fl. Somalia 1: 86 (1993); Paiva in Fontqueria 50: 239, t. 45/f (1998); Gilbert in Fl. Eth. 2, 1: 187, fig. 24.5/6–8 (2000). Type: Sudan, Kordofan, Abu-Gerad, *Kotschy* 8 (G!, holo.; BM!; BREM!; E!; K!; LD!; M!; P!; W!, iso.)

Shrublet or perennial herb, up to 60 cm tall. Stems erect, branched, appressed-pubescent. Leaves alternate, subsessile, linear, oblanceolate to oblong-spathulate, 10–35 × 2–4 mm, obtuse or emarginate, base cuneate, appressed-pubescent. Flowers lilac to brown-reddish, in lax terminal racemes up to 20 cm long, rachis crisped-pubescent; bracts linear, 1 mm long, glabrous, caducous; bracteoles linear, 0.7 mm long, glabrous, caducous; pedicels 2 mm, crisped-pubescent. Posterior sepal ovate, 2–2.2 mm long, ciliate; wing sepals obliquelly-elliptic to ovate, 5–6.5 × 4–4.5 mm, glabrous or pubescent, persistent; anterior sepals 2 mm long, ciliate, free. Upper petals obovate, 3.5–4 × 2–2.2 mm; carina 4.5 × 2.2 mm; crest 0.7 mm long, fimbriate. Stamens 8. Capsule ovate to elliptic in outline, 5–6 × 4–5 mm, asymmetrically emarginate, glabrous, winged, wing 0.5–0.8 mm wide. Seeds conical, 3–3.5 × 1.5–2 mm, sericeous-pubescent, with the convex base completely covered by glands; caruncle pyramidal, up to 0.2 mm long, completely covered by long hairs; appendages 0.2 mm long. Fig. 10: 4, p. 18.

KENYA. Northern Frontier District: Mandera, 15 km S of El Wak on Wajir road, 11 May 1978, *Gilbert & Thulin* 1664!; Meru District: Leopard Rock Camp, 11 June 1963, *Mathenge* 93!; Lamu District: Kiwayu, 23 Oct. 1998, *Luke* 5441!
DISTR. K 1, 4, 7; Senegal, Mauritania, Mali, Nigeria, Chad, Sudan, Ethiopia, Somalia; Egypt, Saudi Arabia, Yemen and Oman
HAB. *Acacia-Commiphora* bushland, bushland on dunes or white sand; 0–500 m

SYN. *Polygala arabica* Boiss., Diagn. Pl. Orient., ser. 1, 1: 9 (1843). Type: Saudi Arabia, near Djedda, *Schimper* 86 (G!, holo.; E!, iso.)
    P. irregularis Boiss. var. *cordofana* Chodat in Mém. Soc. Phys. Hist. Nat. Genève 31 (2): 393 (1893). Type: Saudi Arabia, near Djedda, *Schimper* 861 (G!, lecto.; E!, isolecto.)
    [*P. obtusata auctt., non* DC. (1824)]

44. **Polygala kajii** *Paiva* **sp. nov.**, affinis *P. meonantha* Chodat a qua subfrutex nec herba annua vel perennis, alis 2–2.5 × 1.5 mm nec 3–4 × 1.2–1.6 mm, dense pubescens nec sparse-pubescentibus, capsula applanato-ellipsoidea nec applanato-globosis, 4–5 × 2.5–3 mm nec 2.5–3.5 × 2.5–3.5 mm, dense pubescens nec sparse-pubescentibus, differt. Typus: Tanzania, Mkomazi, Kilevi Bridge, *Abdallah, Mboya et Vollesen* 96/190 (K!, holo.; NHT!, P!, iso)

Subfrutex usque ad 20 cm. Caulis ramosus, 15–20 cm longus, dense crispato-pubescens. Folia alterna, breviter petiolata (petiolo 0.5–1 mm), lamina obovata-elliptica vel obcordata, 5–10 × 2–6 mm, apice emarginata, basi cuneata, dense pubescens. Flores lilacinei, pedicellati, pedicello usque ad 1 mm longo, dense pubescens, in racemos pseudolaterales, paucifloros [1–2(3) flores], 4.5–6 mm longos, dispositi, rachidi dense pubescent; bractea linear, 0.2–0.3 mm longa, caduca; brateolae linearis, 0.25 mm longis, caducae. Sepalum posticum ellipticum, 1.2–1.5 mm longum, pubescens; alae obovato-ellipticae, 4–5 × 1.7–2 mm, dense pubescentia; sepala anteriora libera, linear-ellipticae, 0.7–1 mm longae, pubescentia. Petala superiora oblonga-elliptica, 2–2.5 × 1.5 mm; carina 3–3.2 × 1.5 mm, cristata, crista minuta, 0.3–0.5 mm longa, ramosa. Stamina octo. Capsula applanato-ellipsoidea, 4–5 × 2.5–3 mm, bilobata, alata (ala 0.2–0.3 mm), dense pubescens et ciliata. Semina ovoidea, 3–4 × 1.5–1.8 mm, sericio-pubescentia, carunculata; caruncula subsymmetrica, appendicibus adpressis praedita.

Dwarf-shrublet or woody-based perennial herb up to 20(–40) cm tall, from a woody rootstock 15–20 cm long. Stems slender, much branched from the base, densely pubescent. Leaves alternate, shortly petiolate (petiole 0.5–1 mm), obovate-elliptic to obcordate, 5–10(–15) × 2–6(–6.5) mm, emarginate to obtuse, base cuneate, densely

FIG. 15. *Polygala kajii* – **1**, habit × ²/₃; **2**, flowering shoot × 2; **3**, flower × 8; **4**, stamens and style × 12; **5**, capsule ad sepals × 6; **6**, seed × 12. 1, 3, 4 from *Greenway & Kanuri* 13069; 2, 5, 6 from *Abdullah et al.* 96/190. Drawn by Juliet Williamson.

pubescent on both sides. Flowers purple with pale green veins, in short lateral 1–2(–4)-flowered racemes 4.5–6 mm long, rachis densely pubescent; bracts linear, 0.2–3 mm long, pubescent, caducous; bracteoles linear, 0.2 mm long, pubescent, caducous; pedicels up to 1(–1.5) mm, densely pubescent. Posterior sepal elliptic, 1.2–1.5 mm long, pubescent; wing sepals obliquely obovate-elliptic, 4–5(–6) × 1.7–2(–2.5) mm, densely pubescent, with conspicuous veins; anterior sepals linear-elliptic, 0.7–1 mm long, pubescent, free. Upper petals oblong-elliptic, 2–2.5 × 1.5 mm; carina 3–3.2 × 1.5 mm; crest very short, 0.3–0.5 mm long, fimbriate. Stamens 8. Capsule elliptic in outline, 4–5(–6) × 2.5–3 mm, bilobate, densely pubescent and ciliolate on the margin; narrowly winged, wing 0.2–0.3 mm wide. Seeds ovoid, 3–4 × 1.5–1.8 mm, sericeous (hairs exceeding the base of the seed); caruncle subsymmetric, with three appendages up to near the base of the seed 1.5–2 mm long. Fig. 15, p. 53.

KENYA. Teita District: Tsavo National Park, Voi Gate–Lugard Falls road, 20 Jan. 1967, *Greenway & Kanuri* 13069!; and Ndara to Voi, 29 Nov. 1994, *Luke et al.* 4248
TANZANIA. Pare District: Mkomazi Game Reserve, Kilevi Bridge, 12 June 1996, *Abdallah, Mboya & Vollesen* 96/190!
DISTR. **K** 7; **T** 3; Somalia
HAB. Open to closed bushland; 350–450 m

SYN. *P. erlangeri sensu* Thulin, Fl. Somal. 84 (1993) *pro parte, non* Gürke (1912)

NOTE. This species is named after Kaj Vollesen, one of the botanists who knows the East African flora extremely well and who was one of the collectors of the type specimen.

45. **Polygala meonantha** *Chodat* in E.J. 48(1–2): 326 (1912); Paiva in Fontqueria 50: 174 (1998); Thulin, Fl. Somalia 1: 84 (1993); Gilbert in Fl. Eth. 2, 1: 182, fig. 24.2/10–12 (2000). Type: Ethiopia, Arussa-Galla, *Ellenbeck* 2055 (B†, syn.); Taro-Gumbi, *Ellenbeck* 2048 (B†, syn.); Ethiopia, Sidamo, 38 km from Negele, on Filtu road, *Gilbert* 3358 (K!, neo.)

Annual or short-lived perennial sprawling herb 5–20 cm tall. Stems slender, branched, crisped-pubescent. Leaves alternate, petiolate (petiole 0.5–1 mm, crisped-pubescent), linear to lanceolate, (10–)20–35(–40) × (0.5–)2–5 mm, acute to apiculate, base cuneate, secondary venation and reticulation inconspicuous, sparsely pubescent. Flowers pale pink, in 1–5-flowered lateral racemes 0.7–2.5 cm long, rachis crisped-pubescent; bracts ovate-lanceolate, 0.5–0.8 mm long, crisped-pubescent, soon caducous; bracteoles ovate-lanceolate, 0.5 mm long, crisped-pubescent, soon caducous; pedicels (0.5–)1–2 mm, crisped-pubescent. Posterior sepal lanceolate, 1.7–2 × 0.8–1 mm, acute, crisped-pubescent and ciliolate; wing sepals obliquely elliptic, 3–4 × 1.2–1.6 mm, usually apiculate, sparsely pubescent; anterior sepals lanceolate, 1.5 × 0.7 mm, acute, crisped-pubescent and ciliolate, free. Upper petals obovate, 2.5–3 × 1.1.2 mm; carina 3.5–4.5 × 1.5–2 mm; crest fimbriate. Stamens 8. Capsule ovate-orbicular in outline, 2.5–3.5 mm in diameter, asymmetrically bilobate, sparsely pubescent, very narrowly winged, wing 0.1–0.2 mm wide. Seeds ovoid, 2–2.8 × 1.5 mm, sericeous, caruncle 0.2 mm long, with three short appendages.

KENYA. Northern Frontier District: 30 km on Ramu–Malka Mari road, 6 May 1978, *Gilbert & Thulin* 1529!; Meru District: 1 km outside Meru National Park, S of Ntoe Ranger's Post, 23 Dec. 1969, *Gillett* 18885!; Kwale District: 10 km from Taru on Mombasa road, 20 Nov. 1992, *Harvey, Mwachala & Vollesen* 59!
DISTR. **K** 1, 4, 7; Ethiopia, Somalia
HAB. Scattered tree grassland, *Acacia* bushland, woodland; 0–700 m

46. **Polygala transvaalensis** *Chodat* in Mém. Soc. Phys. Hist. Nat. Genève 31 (2): 374, tab. 29/18 (1893); Paiva in Fontqueria 50: 245, t. 46/c (1998). Type: South Africa, Transvaal, Apies Poort, *Rehmann* 4198 (K!, lecto.)

Perennial herb, 10–20 cm tall. Stems slender, erect, branched, crisped-pubescent. Leaves alternate, subsessile, narrowly-elliptic to linear, 10–30 × (2–)3–11 mm, acute to rounded and mucronate, base cuneate, secondary venation and reticulation inconspicuous, minutely pubescent to glabrous. Flowers pink, crimson, bluish or purple, in 3–6-flowered lateral racemes 1–4 cm long, rachis crisped-pubescent; bracts lanceolate, 0.5–0.8 mm long, crisped-pubescent, caducous; bracteoles lanceolate, 0.5 mm long, crisped-pubescent, caducous; pedicels 2–3.5 mm, crisped-pubescent. Posterior sepal ovate, 2–3 × 1.5–1.8 mm, acuminate, ciliolate; wing sepals greenish, obliquely elliptic, 4.5–6 × 2.5–3 mm, sparsely pubescent to glabrous; anterior sepals lanceolate, 1.5–2 × 0.7–1.3 mm, acute, ciliolate, free. Upper petals obovate, 4–5 × 2.5–3.5 mm; carina 4–5 × 1.5–3 mm; crest 1.5–2.5 mm long, fimbriate. Stamens 8. Capsule suborbicular in outline, 4.5–6 mm in diameter, asymmetrically bilobate, sparsely pubescent, narrowly winged, wing 0.5 mm wide. Seeds ovoid, 2.5–3 × 1.3–1.5 mm, sericeous-pubescent, caruncle 0.2 mm long, with three short appendages. Fig. 10: 5, p. 18.

subsp. **kagerensis** (*Lebrun & Taton*) *Paiva* in Fontqueria 50: 246, t. 46/d (1998). Type: Rwanda, Gabiro, *Lebrun* 9533 (BR!, holo.; K!, iso.)

Perennial prostrate herb or subshrub 5–15 cm tall; leaves oblong-obovate to narrowly elliptic, (5–)10–30 × (2–)3–11 mm, acute to rounded and mucronate; flowers in 1–6-flowered lateral racemes 1–2 cm long; wing sepals greenish, obliquely elliptic, mucronate, 4–5 × 2.5–3 mm; seeds ovoid, 2.5 × 1.2 mm, sericeous-pubescent.

Uganda. Ankole District: Rwampara, 20 Mar. 1938, *Snowden* 1664!; Masaka District: Nabugabo Lake, 10 Oct. 1953, *Drummond & Hemsley* 4732!
Kenya. Masai District: Narok, top of Isuria Escarpment, on Narok–Kilgoris road, 15 Apr. 1961, *Glover, Gwynne & Samuel* 525B!
Tanzania. Lushoto District: Mtai–Malindi road, near Kidologwai, 19 May 1953, *Drummond & Hemsley* 2648!; Morogoro District: Nguru Mts, near Maskati Mission, 10 June 1978, *Thulin & Mhoro* 3114!; Lindi District: Lutamba, 29 Jan. 1935, *Schlieben* 5927!
Distr. **U** 2, 4; **K** 6; **T** 1, 3, 6, 8; Rwanda
Hab. Scattered tree grassland or grassland; 400–2000 m

Syn. *Polygala kagerensis* Lebrun & Taton in Expl. Parc Nat. Kagera, Miss. J. Lebrun 1937–1938, Contr. État Fl. Parc Nat. Kagera: 81 (1948); Petit in F.C.B. 7: 267, fig. 7/G (1958)

Note. *P. transvaalensis* has two subspecies: the southern subsp. *transvaalensis* [Zimbabwe, South Africa (Trasvaal, Orange and Natal) and Lesotho] and the central/east subsp. *kagerensis*. The latter is a sprawling perennial herb or subshrublet (5–15 cm tall), with shorter, mucronate wing sepals (4–5 mm long) and smaller seeds (2.5 × 1.2 mm); while the former is an ± erect, perennial herb (10–20 cm tall), with longer, not mucronate (apiculate), wing sepals (4.5–6 mm long) and larger seeds (2.5–3 × 1.3–1.5 mm).

47. **Polygala goetzei** *Gürke* in E.J. 28: 417 (1900); T.T.C.L.: 457 (1949); Exell in F.Z. 1, 1: 312 (1960); Paiva in Fontqueria 50: 246, t. 46/h (1998). Type: Tanzania, Kilosa District: Luhembe, *Goetze* 406 (B†, holo.; BM!, lecto.)

Perennial, erect herb, 30–40 cm tall. Stems slender, branched, crisped-pubescent. Leaves alternate, subsessile, elliptic to narrowly elliptic, 10–70 × 5–25 mm, rounded to acute and acuminate, base cuneate, with the secondary venation and reticulation conspicuous, sparsely crisped-pubescent to glabrescent. Flowers greenish, in 3–15-flowered lateral racemes 1.5–4 cm long, rachis crisped-pubescent; bracts lanceolate, 0.7 mm long, crisped-pubescent, caducous; bracteoles lanceolate, 0.5 mm long, crisped-pubescent, caducous; pedicels 1.5–3 mm, crisped-pubescent. Posterior sepal ovate, 2.5–2.8 × 1.2 mm, ciliolate; wing sepals greenish, obliquely elliptic, apiculate, 4–5 × 2.5–3 mm, sparsely pubescent to glabrous; anterior sepals lanceolate, 2 × 0.7–1 mm, ciliolate, free. Upper petals broadly obovate and shortly unguiculate, 3.5 × 3.5–3.2 mm; carina 3 × 2.5 mm; crest 1.5 mm long, fimbriate. Stamens 8. Capsule broadly elliptic in outline, 4 × 5 mm, glabrous, narrowly winged, wing 0.5–0.8 mm wide. Seeds broadly ovoid, 2.5 × 2 mm, sericeous-pubescent, caruncle 0.3 mm long, appendages up to half as long as the seed (1.2 mm long). Fig. 10: 6, p. 18.

TANZANIA. Ulanga District: Mahenge, 21 Mar. 1932, *Schlieben* 1943!; Lindi District: Rondo Plateau, W edge of Rondo Forest Reserve, 19 Feb. 1991, *Bidgood, Adlallah & Vollesen* 1652!
DISTR. **T** 6, 8; Zambia, Malawi and Mozambique
HAB. Open woodland and bushland; 450–1000 m

SYN. *Polygala goetzei* Gürke var. *depauperata* Chodat in E.J. 48: 326 (1912). Type: Mozambique, near Beira, *Schlechter* 12248 (B†, holo.; PRE!, lecto.)

48. **Polygala virgata** *Thunb.*, Prodr. Fl. Cap. 2: 120 (1800); T.T.C.L.: 455 (1949); Petit in F.C.B. 7: 266 (1958); Exell in F.Z. 1, 1: 318 (1960); Cribb & Leedal, Mountain Fl. S. Tanz.: 62, t. 10D (1982); Paiva in Fontqueria 50: 249, t. 12/1 (1998). Type: South Africa, Cape Province, *Thunberg* s.n. (UPS!, holo.)

Shrub or shrublet up to 3 m tall. Stems virgately branched, glabrous or nearly so. Leaves alternate, shortly petiolate (petiole 1 mm), linear, narrowly elliptic to oblong-lanceolate, 20–90 × 1.5–5 mm, acute, base cuneate, pubescent when young, glabrous later. Flowers deep purple to pale lilac (rarely with wing sepals whitish, but, if so, with pink veins), in terminal many-flowered racemes up to 20 cm long; rachis sparsely pubescent to glabrous; bracts linear, 2.5–3 mm long, caducous; bracteoles linear, 2 mm long, caducous; pedicels up to 10 mm, pubescent. Posterior sepal keel-shaped, 4–6 mm long, ciliate; wing sepals suborbicular or ± oblong-elliptic, (8–)10–17 × (6–)9–15 mm, glabrous; anterior sepals lanceolate, 3.5–5 × 3 mm, ciliate, free. Upper petals obovate-spathulate, 5–8 × 3–4 mm; carina 9–13 × 6–8 mm, ciliate; crest 4–6 mm long, fimbriate. Stamens 8. Capsule obliquely obovate-elliptic in outline, 7.5–10 × 5–8 mm, winged, wing 1 mm wide, glabrous. Seeds ellipsoid-cylindric, (3.7–)4–4.5 × 1.5 mm, silky-pubescent; caruncle asymmetric, 1–1.3 mm long, with almost inconspicuous appendages, 0.1–0.2 mm long.

var. **decora** (*Sond.*) *Harv.* in Harv. & Sond., Fl. Cap. 1: 85 (1860); Exell in F.Z. 1, 1: 318, t. 56/14, t. 58/A (1960); Paiva in Fontqueria 50: 250, tab. 6/a; tab. 48/a; tab. 49/k, l; (1998). Type: South Africa, Port Natal, *Gueinzius* 23 (?S, holo.; W!, iso.)

Stems densely pubescent, rarely sparsely pubescent; leaves 20–90 × 5–20 mm; seeds ellipsoid-cylindric, (3.7–)4–4.5 × 1.5 mm, soft-pubescent, with short hairs, which do not surpass the base of the seed. Fig. 11: 1, p. 20

TANZANIA. Lushoto District: Mbwei–Mlola, 20 Feb. 1987, *Kisena* 545!; Rungwe District: Pangundutani, Sawago, 25 Oct. 1947, *Brenan & Greenway* 8205!; Songea District: Matengo Hills, Lupembe Hill, 20 May 1956, *Milne-Redhead & Taylor* 10257!
DISTR. **T** 3, 4, 7, 8; Congo-Kinshasa, Zambia, Malawi, Mozambique, Zimbabwe, Swaziland, Lesotho and South Africa
HAB. Forest edges, secondary bushland derived from forest and upland grassland; 900–2600 m

SYN. *P. decora* Sond. in Linnaea 23: 14 (1850)
    P. ourolopha Chodat in E.J. 48 (1–2): 331 (1912). Type: Mozambique, Gorongosa, *Carvalho* s.n. (B†, holo.; COI!, lecto.)
    [*P. virgata sensu* Brenan & Greenway, T.T.C.L.: 455 (1949), *non* Thunb. (1800)]

NOTE. *P. virgata* has a large distribution, from Congo-Kinshasa and Tanzania in the North, to the Cape in the South, and it is extraordinary polymorphic. So a lot of varieties and even species were described, the majority of them not satisfactorily distinguished from the type variety. Some authors consider that the species is best treated as merely variable and that it is not useful to try to separate it into varieties. But the tropical African taxon, which occurs from Transvaal and Natal to the North to Tanzania and Congo-Kinshasa is really a distinct variety, var. *decora* (Sond.) Harv. This has stems sparsely to densely pubescent, leaves broader (5–20 mm wide), seeds with short hairs, which do not surpass the base of the seed, while the other varieties have stems glabrous or nearly so, leaves narrower (up to 5 mm wide) and seeds with long hairs, which surpass the base of the seed. In South Africa this variety has forms intermediate to the typical variety, var. *virgata*. In the Southeast of Cape Province occurs another distinguishable variety, var. *speciosa* (Sims) Harv.

The specimen *Archbold* 2445, from Milo (**T** 7, Njombe District) has whitish wing sepals with pink veins. The two species of subsection *Megatropis* with white sepals and similar to *P. virgata* var. *decora*, are *P. wittei* Exell, from Congo-Kinshasa (Haut-Katanga) and *P. lactiflora* Paiva & Brummitt [from Malawi (Nika Plateau)]. But these two species have smaller flowers (wing sepals 8–11 × 5–7 mm; *P. virgata* var. *decora* (8–)10–17 × (6–)9–15 mm) and glabrous staminal tube and style, while in *P. virgata* var. *decora* they are ciliolate.

49. **Polygala abyssinica** *Fresen.* in Mus. Senckenb. 2: 273 (1837); Blundell, Wild Fl. E Africa t. 717 (1978); Thulin, Fl. Somalia 1: 86 (1993); U.K.W.F. ed. 2: 55 (1994); Paiva in Fontqueria 50: 258, t. 48/g-i (1998); Gilbert in Fl. Eth. 2, 1: 185, t. 24.4/3–5 (2000). Type: Ethiopia, Semien, *Rüppell* 38 (FR!, holo.; COI!, iso.)

Perennial herb, 30–50 cm tall. Stems erect, ± branched from the woody base, striate, sparsely pubescent to glabrous. Leaves alternate, shortly petiolate (petiole 0.5–1 mm), linear to oblong-lanceolate, 1–40 × 0.5–4(–6) mm, acute to almost obtuse, minutely hairy to glabrous. Flowers purple or pinkish, in elongate terminal racemes up to 20 cm long; rachis sparsely pubescent to glabrous; bracts linear, 1.5–3 mm long, caducous; bracteoles linear, 1–1.5 mm long, caducous; pedicels 2.5–3.5 mm, glabrous. Posterior sepal lanceolate, 2.5–3.5 mm long glabrous; wing sepals elliptic to slightly obovate, 7–8.5 × 3.5–5 mm, with prominent purple or greenish veins, apiculate, glabrous; anterior sepals linear-lanceolate, 2–3 mm long, glabrous, free. Upper petals oblong-elliptic to spathulate, 3.5–5 × 2 mm; carina 5–6.5 × 3 mm; crest 1.5–2 mm long. Stamens 8. Capsule obliquely obovate-elliptic in outline, 5–5.5(–6) × 3–3.5 mm, very narrowly winged, glabrous. Seeds ellipsoid, 4–4.5 × 1.5 mm, with dense white, silky hairs; caruncle asymmetric, 0.3–0.5 mm long, with almost inconspicuous appendages, 0.1–0.2 mm long. Fig. 11: 2, p. 20.

UGANDA. Karamoja District: Morongole Mts, 11 Nov. 1939, *Thomas* 3314!
KENYA. Naivasha District: Longonot, 29 Aug. 1930, *Napier* 198!; North Nyeri District: 50 km on Nanyuki–Isiolo road, 27 Mar. 1978, *Gilbert* 5039!; Masai District: 11 km from Olulunga, 6 June 1961, *Glover, Gwynne & Samuel* 1652!
TANZANIA. Masai District: Ol Doinyo Sambu Mt, 18 Jan. 1936, *Greenway* 4416!
DISTR. **U** 1; **K** 3, 4, 6; **T** 2; Sudan, Ethiopia, Djibouti and Somalia; Arabian Peninsula
HAB. Scattered tree grassland, grassland, in rock crevices, often on volcanic ash or black soils; 1000–2700 m

SYN. *Polygala abyssinica* Fresen. var. *adoensis* Chodat in Mém. Soc. Phys. Hist. Nat. Genève 31 (2): 389 (1893). Type: Ethiopia, near Adoa, *Schimper* s.n. (B†, holo.; K!, OXF!, W!, iso.)
    P. abyssinica Fresen. var. *parviflora* Chodat in Mém. Soc. Phys. Hist. Nat. Genève 31 (2): 389 (1893). Type: Ethiopia, Lake Amba, *Schimper* 445 (B†, holo.; BM!, iso.)
    P. abyssinica Fresen. var. *pilosa* Chiov. in Nuovo Giorn Bot. Ital., Sér. 2, 26(2): 92 (1919), *nom. nud.*, based on: Ethiopia, Uogherico, *Beccari* s.n. (FT!)

NOTE. I did not find the type [Kenya, Elmenteita, 1500–2100 m, *Scott Elliot* 6689 (B†?, holo.)] of *P. alata* Chodat [in Engl. E.J. 34: 200 (1896); Paiva in Fontqueria 50: 307 (1998)], but there is at Kew one specimen of *Scott Elliot* 6689 from Lake Nakuru (in the same district as Elmenteita) handwritten as *P. alata* Chodat, which is *P. abyssinica* Fresen. Even Chodat [in Bull. Herb. Boisier 4 (12): 910 (1896)] says that his species was similar to *P. abyssinica* Fresen. Later [in Engl., E.J. 48 (1–2): 327 (1912)], he cites again *P. alata* Chodat from East Africa [Tanzania, Longido-Berge (Uhlig in Exped. D. Otto-Wintes-Stiftung n. 199]. I have not found those specimens or their duplicates, but I am almost sure that they belong to *P. abyssinica*, which is very polymorphic and common in Nakuru District (**K** 3) and Masai District (**T** 2).

50. **Polygala steudneri** *Chodat* in Mém. Soc. Phys. Hist. Nat. Genève 31 (2): 390 (1893); Paiva in Fontqueria 50: 261, t. 50/f (1998); Gilbert in Fl. Eth. 2, 1: 185, t. 24.4/1–2 (2000). Type: Ethiopia, Semien, Mt Buahit, *Steudner* s.n. (B!, holo.)

Perennial herb, up to 18 cm tall. Stems procumbent, slender, spreading from a woody rootstock, much branched, crisped-pubescent. Leaves alternate, shortly petiolate (petiole 0.5–0.8 mm, narrowly winged, crisped-pubescent), linear-elliptic, 7–12 × 1.5–3.5 mm, rounded to obtuse, mucronate, slightly base cuneate, crisped-pubescent on the margins and main vein on both sides. Flowers mauve or pink, in terminal (and rarely lateral) racemes 2–4 cm long; rachis crisped-pubescent; bracts linear, 0.7–1 mm long, glabrous, caducous; bracteoles linear, 0.5 mm long, glabrous, caducous; pedicels 1.5–1.8 mm, glabrous. Posterior sepal lanceolate, 2.7–3 mm long, glabrous; wing sepals elliptic, 5–5.5 × 3–3.5 mm, with a conspicuous middle stripe, glabrous; anterior sepals linear-lanceolate, 2.5–2.8 mm long, glabrous, free. Upper petals oblong-elliptic, 3–3.5 × 1.5 mm; carina 5 × 2.5 mm; crest 2.5 mm long. Stamens 8. Capsule elliptic in outline, 3–4 × 2.5 mm, very narrowly winged (wing 0.1–0.2 mm wide), glabrous. Seeds cylindric, 2.5 × 1 mm, with dense white silky hairs; caruncle asymmetric, 0.3 mm long, appendages 1.5 mm long, up to the half length of the seed. Fig. 11: 3, p. 20.

UGANDA. Mt Elgon, Jan. 1918, *Dümmer* 3365!
KENYA. Mt Elgon, July 1933, *Dale* 3101! & Mt Elgon, Kimilili Trail, 26 Nov. 1997, *Wesche* 1827!; South Nyeri District: Cave Waterfall, 21 June 1962, *Coe* 775!
TANZANIA. Mbulu District: Mt Hanang, NE slope, 8 Feb. 1946, *Greenway* 7668!
DISTR. **U** 3; **K** 3–5; **T** 2; Ethiopia; Saudi Arabia, Yemen
HAB. Afroalpine grassland, grassland, often in rocky places; (2550–)3000–4050 m

SYN. *Polygala negri* Chiov. in Ann. Bot (Roma) 9(3): 315 (1911). Type: Ethiopia, Mt Menagesha [Mangascia], *Negri* 401 (FT!, syn.); Ethiopia, Gabba, Mt Wechacha [Uaciaccia], *Negri* 424 (FT!, syn.)

INDEX TO POLYGALACEAE

59

New names validated in this part

Polygala kajii *Paiva*

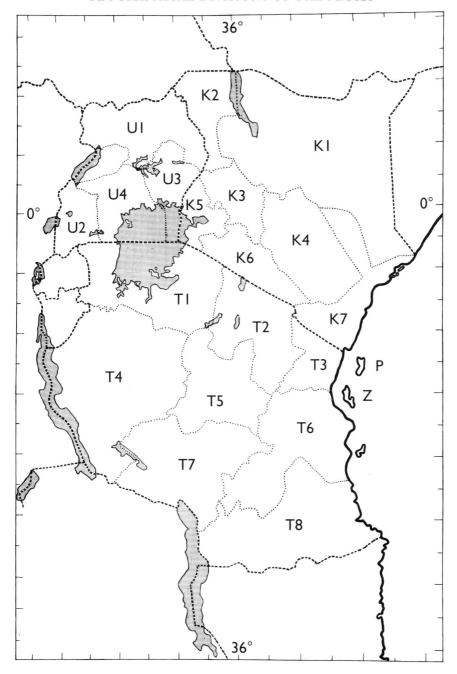

PLANTS PEOPLE
POSSIBILITIES

First published in 2007 by
Royal Botanic Gardens, Kew
Richmond, Surrey, TW9 3AB, UK
www.kew.org

ISBN 978 1 84246 191 4

British Library Cataloguing in Publication Data
A catalogue record for this book is available from the British Library

Design and typesetting by Margaret Newman,
Kew Publishing, Royal Botanic Gardens, Kew.

Printed in the UK by Hobbs the Printers

For information or to purchase all Kew titles please visit
www.kewbooks.com or email publishing@kew.org

All proceeds go to support Kew's work in saving the world's plants for life